MIMESIS
INTERNATIONAL

AF705661

SHIFT / PHILOSOPHICAL SERIES
n. 3

Directed by: Daniela Calabrò

International Editorial Board:
Renaud Barbaras (Sorbonne – Paris IV), Greg Bird (Wilfrid Laurier University), Valerio Cappozzo (University of Mississippi), Danielle Cohen-Levinas (Sorbonne – Paris IV), James Connelly (University of Hull), Marc Crépon (ENS – Paris), Georges Didi-Huberman (EHESS – Paris), Francesco Saverio Festa † (University of Salerno), Alfonso Galindo Hervàs (University of Murcia), Dario Giugliano (Academy of Fine Arts – Naples) Evelyne Grossman (University of Paris VII), Yvonne Hütter (University of München), Rahel Jaeggi (University of Humboldt – Berlin), Laurens ten Kate (University of Utrecht), Jérôme Lebre (University of Strasbourg), Enrica Lisciani-Petrini (University of Salerno), Boyan Manchev (University of the Arts – Berlin), Aichä Messina (University of San Diego Portales), Ginette Michaud (University of Montréal), Jean-Luc Nancy † (University of Strasbourg), Stefan Nowotny (Goldsmyths College – University of London), Anne O'Byrne (Stony Brook University – NY), Laura Odello (Brown University), Rosalia Peluso (University "Federico II" – Naples), Frans van Peperstraten (Tilburg University –Netherlands), Francesco Piro (University of Salerno), Andrea Potestà (University "Pontificia Catolica" – Santiago), Jacques Rancière (University Saint-Denise – Paris VIII), Jacob Rogozinski (University of Strasbourg), Aukje van Rooden (University of Amsterdam), Olivier Tonneau (University of Cambridge), Marcia Sá Cavalcante Schuback (Södertörn University), Hent de Vries (Johns Hopkins University – Baltimore), Erik Wallrup (Stockholm University).

Daniela Calabrò

A DIASTOLE WITHOUT SYSTOLE

Merleau-Ponty's New Humanism

MIMESIS
INTERNATIONAL

The publication of this book was made possible by Funding for Basic Research (FARB) from the Department of Human Sciences, Philosophy and Education of the University of Salerno.

This volume has undergone a double-blind peer review process.

© 2025 – Mimesis International (Milan – Udine)
www.mimesisinternational.com
e-mail: info@mimesisinternational.com

Isbn: 9788869775253
Book series: *SHIFT / Philosophical Series*, n. 3
Series Issn: 3103-4381

© MIM Edizioni Srl
P.I. C.F. 02419370305

ACKNOWLEDGEMENTS

This book is the result of twenty years of research on Maurice Merleau-Ponty.

But it is also the result of encounters, exchanges of ideas, conversations, and relationships.

For all of this, I must thank the friends and colleagues who, in various ways, have been a fruitful point of reference and reflection for me. I therefore feel the duty to mention them here one by one, starting with Enrica Lisciani-Petrini, to whom this work is dedicated.

Here, then, is my *tabula gratulatoria*: Maurizio Cambi, Carmelo Colangelo, Marianna Esposito, Roberto Esposito, Gian Paolo Faella, Rita Fulco, Paolo Fusco, Dario Giugliano, Giulio Goria, Luigi Imperato, Alessio Lembo, Fiorenza Manzo, Paola Martino, Valentina Mascia, Rosalia Peluso, Marco Piasentier, Francesco Piro, Marco Russo, Luca Scafoglio, Cinzia Soddu, Davide Tarizzo, Renato Trombelli, Aldo Trucchio, Massimo Villani, Angelo Maria Vitale.

A heartfelt thank-you to Silvia Ciccarelli and Lucia Valentino for their dedication and generosity of spirit in revising the texts.

Can not be missing from this *tabula gratulatoria*: Girolamo Cotroneo, Francesco Saverio Festa, and Jean-Luc Nancy – in memory of the time that was ours.

Allow me also to express a thought of gratitude for the unfailing support of my wonderful Family.

Finally, my gratitude goes to the "silent presence" that constellates and accompanies the steps of my existence, illuminating them with love.

To Enrica

TABLE OF CONTENTS

PREFACE

"At the conclusion of a reflection which at first isolates him, the philosopher, in order to experience more fully the ties of truth which bind him to the world and history, finds neither the depth of himself nor absolute knowledge, but a renewed image of the world and of himself placed within it among others. His dialectic, or bis ambiguity, is only a way of putting into words what every man knows well the value of those moments when bis life renews itself and continues on, when be gets hold of himself again, and understands himself by passing beyond, when bis private world becomes the common world. These mysteries are in each one of us as in him. What does be say of the relation between the soul and the body, except what is known by all men who make their souls and bodies, their good and their evil, go together in one piece? What does he teach of death, except that it is hidden in life, as the body in the soul, and that it is this understanding, as Montaigne said, which brings 'a peasant and whole peoples to die, just as surely as philosophers?' The philosopher is the man who wakes up and speaks. And man contains silently within himself the paradoxes of philosophy, because to be completely a man, it is necessary to be a little more and a little less than man"[1].

This is the concluding passage of *In Praise of Philosophy and Other Essays*, one of Merleau-Ponty's most significant works. Echoing these meaningful reflections on the sense of human life are the words of Enzo Paci, in a joint note to

1 Merleau-Ponty, M., *In Praise of Philosophy and Other Essays*. Northwestern University Press, Evanston, Ill. 1988, pp. 63-64.

Merleau-Ponty's text: "In the end, philosophy inserts itself into human life and is not only an expression of life – it is internal to life. Every human being therefore carries within himself the very paradoxes for which philosophy is reproached. To strip philosophy of its paradoxes is like denying life itself. A human being is *less* than human insofar as he can never fully be himself, never fully satisfied, never completely realized. He is *more* than human insofar as, precisely because he is never finished, he is always in search of a new and more meaningful sense of life and of history"[2].

The development of an innovative vision of humanism[3] that integrates corporeality and the concrete experience of the individual as a central part of the human being is the foundational core of Merleau-Ponty's philosophy. In contrast with classical humanism, often rooted in an abstract and idealistic conception of man, Merleau-Ponty's humanism is distinguished by its grounding in perception and embodiment. For Merleau-Ponty, the human being is not a mind isolated from the body or from the world, but an incarnate being who knows and understands reality through the body, in a continuous and dynamic relationship with the surrounding environment.

---

2 Paci, E., *Note 59* to Merleau-Ponty, Maurice, *Elogio della filosofia*. Turin: Paravia, 1958, p. 80 (my translation).

3 Humanism has profoundly traversed European philosophical thought, proposing a vision in which the human being stands at the very heart of his relationship with the world. Since the Renaissance, humanism has been the rediscovery of the individual—of his rights and of his capacity to confer meaning upon reality—affirming a stance that places the human at the centre of knowledge and of moral life. Though continually reshaped across the centuries, this orientation has nonetheless retained a decisive role within the philosophical reflections of the twentieth century. Phenomenology and existentialism – through thinkers such as Edmund Husserl, Martin Heidegger, Jean-Paul Sartre, and Maurice Merleau-Ponty – have revisited humanism, opening it to new dimensions grounded in perception, in the lived body, and in subjectivity.

Unlike a purely intellectual humanism, the French philosopher – as we shall see in this work – proposes an "embodied humanism," in which the body is the vehicle of perception and the gateway to experience of the world. With his major work, *Phenomenology of Perception* (1945), Merleau-Ponty explores the indissoluble bond between the body and the perceived world, inaugurating a new reflection on what it means to be human. Sensible experience, lived bodily experience, and direct interaction with the world are, for Merleau-Ponty, the foundation of our knowledge and understanding of existence, opening the way to a form of humanism that challenges the idea of a mind separate from the body and highlights the human being as a "being-in-the-world."

From this arises another significant aspect of Merleau-Ponty's humanism: his vision of subjectivity and intersubjectivity. The human being, according to Merleau-Ponty, is a *Being-in relation*: his identity and self-understanding depend on encounters and interactions with others. In this context, language and communication play a fundamental role, allowing us to build a bridge between the self and the Other, and to construct a shared world. This conception of the human being as a perceiving subject and as a being in relation with others and with the world constitutes a response to the limits of modern thought, which tended to fragment the individual into a purely rational dimension.

Merleau-Ponty's humanism finally assumes a political and historical dimension, as it conceives the human being as an integral part of the social and historical reality in which he lives. In texts such as *Humanism and Terror* (1947), the philosopher explores political commitment and the responsibility of the individual within history, laying the foundations for an ethics of embodied freedom and responsibility. This political vision of humanism stands in opposition to the idea of an isolated subject, promoting instead a collective and solidaristic humanity.

In this work, the main elements of Merleau-Ponty's humanism will be explored: corporeality, perception, intersubjectivity, and the historical and political dimension of his thought. Through these themes, it will become clear how Merleau-Ponty's humanism represents an original and profound response to fundamental questions about the human being, proposing a vision of humanity that integrates body, mind, subject, object, Self, and other, and that reveals a new possibility of inhabiting the world: in a continuous and ever-emerging "diastole without systole."

PART 1
THINKING THE BODY

CHAPTER 1
TWENTIETH-CENTURY SCENARIOS AND PHENOMENOLOGICAL PREMISES

§ 1. *The Crisis of the Sciences, the Impact of Psychoanalysis, the Fragmentation of the Self*

The Twentieth century is marked by the decline of definitive and irrefutable philosophical and scientific systems. Indeed, the very concept of scientific rationality – grounded in the logical nexus of cause and effect – enters into crisis in the wake of certain scientific discoveries. The millenary confidence in the continuity of natural phenomena (*natura non facit saltus*), as well as in the certainty of an objective knowledge of reality, gradually collapses under the impact of those discoveries which, from the first decades of the century, shattered the former physical systems of explanation of the real, all founded upon the postulate or principle of causality.

The radical evolution of mathematics and physics reveals the untenability of the Galilean–Newtonian framework, which had been based upon the causative principle and the supposed objectivity of experimental data.

Science thereby transforms almost into a *doxa*, that is, into a form of knowledge whose theories are not systems of propositions that are true in themselves, but rather systems of hypotheses – a knowledge that is therefore neither certain nor definitive, but conjectural and fallible. This represents, without doubt, a substantial shift in the previously dominant conception of science as *ἐπιστήμη*.

The first significant moment in the transformation of the image of scientific knowledge is constituted by the discovery of non-Euclidean geometries. The new situation that emerged in the field of geometry, with the demonstration of the internal coherence of non-Euclidean systems – and even earlier with their very construction – immediately raises the question of which among these geometric systems can be considered genuinely reliable.

At the threshold of the non-Euclidean revolution, the prevailing perspective on the nature of geometry was that elaborated by Kant in the *Critique of Pure Reason*: geometry is a system of synthetic *a priori* propositions, that is, propositions that describe the real structure of physical space and whose truth is independent of experience. Yet now, confronted with a plurality of geometric systems, each internally coherent (and thus logically equivalent), but mutually incompatible, and with the consequent crisis of confidence in evidence as a criterion of truth, it becomes both legitimate and necessary to reconsider the foundations of geometry.

This gives rise to an important critical reflection on the nature of geometric foundations and on the relationship between geometry and the physical world – a reflection that develops with particular significance into the early decades of the twentieth century.

The common orientation expressed by the various forms of this critical reflection on the foundations of geometry is decidedly anti-Kantian. Gauss was the first to realize that Euclidean geometry is not necessarily the geometry of physical space, and that its truth cannot be guaranteed *a priori*. A further, significant moment in the inquiry into the foundations of geometry and the relation between geometry and the physical world is represented by Poincaré's epistemological reflection. Rejecting both Kantian apriorism and empirical approaches, Poincaré argues for the conventional character of geometric axioms: the criterion for choosing among different systems cannot reside in a notion of truth that is meaningless to speak of, but rather in the greater or lesser convenience or utility from the standpoint of the scientist. The scientist is thus free to adopt the type of geometry he prefers, under the sole condition that the absolute requirement to avoid contradiction is preserved.

Moreover, the profound transformation in the foundations of the investigation of the physical world – marked by the crisis of the mechanistic research program and by the development of relativity and quantum theories – leads, between the late nineteenth and early twentieth centuries, to a radical reconsideration of the nature of physical theories and, more generally, of scientific theories: their logical structure and epistemological status.

In particular, 1927 represents the culmination of the crisis of scientific rationality. Heisenberg and Bohr dealt a severe blow to the constitutive pillars of physical science: the traditional concepts of symmetry, order, reversibility, and continuity are supplanted by notions that are decidedly more unsettling and less reassuring – quantum discontinuity, discreteness, irreversibility, and indeterminacy. Scientific knowledge, grounded in experimental observation, becomes, in principle, impossible: physics can no longer rely on familiar images. The uncertainty relations simply indicate that the two aspects of the principle of causality – the necessity with which effect follows cause and the infallibility of prediction – do not manifest in subatomic phenomena, and therefore in no part of physical reality.

In this manner, the principle of causality, which had appeared to be the most solid pillar of the theoretical constructions of nineteenth-century science, has been shattered – not by theoretical critique, but by the very experimental advances of science itself. The upheaval implicit in this collapse of the principle of causality is so radical that its full consequences have yet to be comprehended.

This explains the reluctance of many contemporary thinkers and scientists to accept this collapse, as well as the attempts made to uphold the principle despite its manifest inapplicability to science. In *The Logic of Scientific Discovery* (1935), certainly one of the fundamental and most representative works of twentieth-century epistemology, Popper acknowledges the failure of the traditional representation of science and writes: "The old scientific ideal of *ἐπιστήμη* – of absolutely certain, demonstrable knowledge – has proved to be an idol. The demand for scientific objectivity makes it inevitable that every scientific assertion must necessarily remain, forever, in the state of a tentative endeavour. Scientific theories are, and remain, hypotheses: they are conjectures (*dóxai*), opposed to indubitable knowledge (*ἐπιστήμη*)"[1].

Scientific indeterminism thus nullifies the traditional assumption of dogmatic thought regarding the uniqueness of truth and opens the field to the perspective of a truth that is mutable, ever to be constituted.

Alongside the historical analysis of the crisis in the foundations of physical and mathematical sciences, it becomes necessary – both for completeness and coherence – to examine the field of the psychological sciences. The psychological landscape of the early twentieth century, alongside a tendency toward specialization and the treatment of micro-problems within experimental research, is characterized by the emergence of new "schools," each defined by a specific conception and practice of psychological or behavioural analysis.

Already in the last decades of the nineteenth century, a particular attitude spread throughout European psychological research, increasingly oriented toward objectivity thanks to the consolidation of the scientific and experimental perspective. Contributing decisively to this conceptual shift was the growing favour accorded to evolutionary theory among psychologists. Thus, the goal of bringing psychology to procedures of intersubjectively controllable research required a modification of Wundt's model, which analysed the fundamental elements of psychic life primarily through introspection.

1 Popper, K., *The Logic of Scientific Discovery*, Routledge, London-New York 2002, p. 280.

With the advent of the behaviourist turn, which became the principal foundation of the new scientific perspective on human behaviour, psychology benefited not only from a revision of methods but also from the already advanced integration of research domains in physiology and animal psychology. Behaviourism adopted, as a general theoretical premise, the possibility of explaining the psychic phenomena of all animal organisms, provided that all references to unverifiable concepts or "entities" were eliminated, introspective and anthropomorphic appeals were abandoned, and explanations were sought solely in observable material, eschewing purely analogical reasoning.

In other words, the possibility of constructing a "psychology without a soul," as objective psychology was then termed, was firmly asserted. Watson was the most convinced interpreter of this new research perspective. He laid the conceptual foundations that would guide all American psychological research, assigning to psychology the task of studying the objective conditions that determine behaviour while eliminating all reference to introspective mental contents. The explanation of human behaviour, like that of animals, was sought in the identification of causal chains of stimuli and responses, regarded as the minimal units of behaviour. These causal chains, in turn, were investigated primarily in the external environment rather than within the organism itself.

Yet, in these same years, alongside the introspectionist and behaviourist models, the phenomenological approach also began to assert itself. In contrast to Watsonian associationism, which emphasized the additive laws that determine the constitution of the object, and in opposition to Wundtian introspectionism, which nearly dissolved the object into the conceptual analysis of states of consciousness, the phenomenological orientation consolidated in continental European psychology through the group of psychologists known as the Berlin School. This approach, broadly defined as phenomenological, foregrounds the subject-object nexus, which had seemed entirely effaced both by associationism and by introspectionism.

This subject-object relation, a foundational pillar of phenomenological reflection, is investigated in the most complex psychic act: perception. Thanks to C. von Ehrenfels, who introduced the notion of "form qualities" (*qualitäten*), the new psychology – universally known as *Gestalt* psychology or *Gestaltpsycologie* – contributed decisively to redirecting the attention of psychologists, especially those oriented toward experimentation, to the necessity of reformulating experimental designs for the study of higher psychic phenomena[2].

Within this conceptual framework, perception is assigned particular importance as a domain through which the dynamic and synthetic character of psychic life can be apprehended in greater detail. Neither the value of the external object (or world) nor that of the subject is absolutized. "Structures" or "gestalten" are identified both in the physical and mental worlds, and between these two domains an overarching condition is sought that would render their interpretation homogeneous. This condition is formulated in the postulate of isomorphism, that is, a correspondence of forms or structures between the physical and the psychic, between the physiological and the mental spheres, according to an explanatory procedure of an analogical type.

This perspective is primarily expressed in two fundamental psychological laws: the law of non-additive formation of the whole, and the law of Prägnanz, according to which the whole can only be understood if it is not considered as the mere sum, through progressive addition, of primitive elements. These elements cease to be mere addends and become factors subsumed under the character of belonging to the whole. In

---

2 For further information on Gestalt theory, see, to name but a few, Köhler, W., *The Task of Gestalt Psychology*, Princeton University Press 1969, and Koffka, K., *Principles of Gestalt Psychology*, London, Routledge 1999.

this way, the elements that enter sensation possess not only a "structured" function but also a "structuring" function, insofar as they constitute the material and simultaneously contribute to determining its general structure or configuration. The perceptual act is thus regarded as primitive and immediate (in the sense of non-intellectual and non-reflective), objective (in that it is tied to conditions external to the perceiver), and global and unified (in that it is not a mere punctual excitation).

Hence, the understanding of perception as constitutive of the object relies less on associative learning and more on the very structure of the perceptual event: it is regarded as a primary fact, both structured and structuring[3]. According to Gestalt psychology, objects are apprehended or perceived immediately within organized, structured, and dynamic totalities. The decisive contribution of Gestalt theory lies in overcoming the traditional positions of psychophysics, which reduces the subject-object relation to mere mechanism, and of introspectionism, which situates perceptual phenomena solely within the conscious sphere of the individual.

Indeed, the totalizing character of experience, the part-whole reciprocal dynamics, the intrinsic relationality of experiential parts, and the field that structures the entirety of things are elements that characterize the Gestalt methodology, according to which perceptual experience presents itself as a complete whole already imbued with meaning – a profile, so to speak. Perceptual experience is always "meaningful" within a perceptual field (a horizon of sense), without which things would have no significance.

Alongside these fundamental innovations, another decisive development in psychology must be noted: the emergence of psychoanalysis and the discovery of the unconscious. Among all the classical schools of psychology, psychoanalysis, inseparably linked to its founder Sigmund Freud, exerted

---

3 See, in particular, "Two boys play badminton" in Wertheimer, M., *Productive Thinking*, Cham, Springer, 2020, pp. 169 and sub.

the greatest influence on contemporary culture, shaping not only psychology but also literature, sociology, cultural anthropology, education, and philosophy. For this reason, the "psychoanalytic revolution," initially conceived as a method for treating certain mental illnesses, ultimately opened new horizons for the interpretation of human existence.

"Freud", as Moravia notes, "gained a new vantage point from which to observe humanity: the *Es*. With this German term, corresponding to the neuter pronoun "it" (Latin *Id*), he designates the dark, inaccessible part of our personality. It is unconscious, yet tends to emerge at the conscious level, becoming the *Ego*, that is, the portion of the *Es* modified by proximity to and influence of the external world: conscious and unconscious are thus intimately intertwined processes, transforming reciprocally into one another"[4].

Psychoanalysis, therefore, introduces a novel approach into psychology, whereby the understanding of conscious human life is subordinated to the understanding of its unconscious psychic life. As will be seen throughout this work, the findings of these psychological sciences will prove fundamentally important in shaping the thought of Maurice Merleau-Ponty.

§ 2. *Epoché and the Lifeworld: Husserl's Debt*

In the philosophical field, as was mentioned at the outset, the response to the crisis of the sciences is provided first and foremost (and, for our present concerns, most directly) by the advent of phenomenology. To speak of phenomenology is to speak of Edmund Husserl, who proposes the idea of a universal knowledge and of philosophy as a "rigorous science" (*Philosophie als Strenge Wissenschaft*). Yet, according to

---

4 Moravia, S. *Introduzione* a Freud, S. *Filosofia e Psicoanalisi*, edited by S. Moravia (it. transl.), La Nuova Italia, Florence 1981, pp. XIX-XXIII. For further study please refer to Freud's entire corpus of works.

Husserl, the necessity of fulfilling the idea of a rigorous philosophy implies the problem of seeking a universally valid ground or principle; in other words, the requirement to begin becomes primary. The only genuine beginning consists in the givenness of things, for they present themselves to us in any case; they manifest themselves prior to any reasoning, judgment, or decision. Thus, we cannot reject anything but must take note of all that is given.

For Husserl, the human situation is understood as the horizon of manifestation – the manifestativity – of things, which may rightly be called phenomena. Phenomena, then, are things insofar as they are given, insofar as they manifest themselves; or rather, they manifest themselves to us. The horizon of our living is always constituted by a horizon of phenomena brought to evidence. This evidence is the principle that implicitly guides the Husserlian phenomenological perspective from within. The evidence in question expresses, most immediately and simply, the givenness of things in the temporal flow of our perceptual life, oriented outward (toward the world) as well as inward (toward the Self), and precisely for this reason, such evidence precedes any judgment or decision, precedes the exact or empirical sciences, precedes every theory or notion. In this case, the contents of evidence – things, phenomena, or data – reveal an original right, as Husserl asserts in §24 of *Ideen* I and §34 of the *Krisis*: *the original right of all givenness to be accepted simply as it gives itself out to be, but also only within the limits in which it is given there*[5].

It is clear that this right does not pertain to any decision of mine or another's: I do not decide upon this right, and it does

5 Husserl, E., *Ideas Pertaining to a Pure Phenomenology and to a Phenomenological Philosophy*. 1913. Dordrecht, Kluwer Academic Pub, 1983 (p.44); and Husserl, Edmund. *The Crisis of European Sciences and Transcendental Phenomenology: An Introduction to Phenomenological Philosophy*; Translated, with an Introduction, by David Carr. Translated by David Carr, Northwestern University Press, Evanston, Ill. 1970.

not depend upon my recognition. Rather, insofar as I choose not to reject it, I simply inaugurate an authentic consideration of my mode of existence, which is intimately bound to the right of givenness. Where I to attempt its rejection, I would fall into the absurdity of having to acknowledge it, in advance, merely in order to reject it; moreover, this abstract refusal could not prevent my continued living within the horizon of the manifestativity of phenomena and their original right to manifest themselves.

From this, it follows that the principle of evidence is not properly a "principle" if the word is understood equivalently to "axiom." It is a principle, however, if one attends to the more semantically congruent meaning: principle in the sense of "from what to begin," of initiating any consideration, of where to take the first step. To clarify Husserlian reflection on evidence, several moments may be outlined:

1. To move from evidence is to begin with what is given prior to any question, problem, or inquiry.
2. To remain on the ground of evidence is to adhere to the mode *with which things are given.*
3. To remain on the ground of evidence and its original givenness, one must take up what is given within its mode of being given; that is, one must assume it within the limits in which it is actually given.
4. That which is assumed not as it is given, but under *presumption* of reality, is not assumed from evidence and is therefore not truly assumed at all; it is only supposed.
5. Certainly, even the supposed is given, but precisely in the mode of supposition, which has its own evidences. Yet evidence in the mode of supposition is not evidence in the mode of perception, and the supposed must never substitute for the perceived; it must never be assumed as if it were perceived.

From these five points, it becomes clear that reflecting on the ground of evidence entails preventing oneself from *what is given differently from how it is given*; it entails returning to the lived present, to our presence, that is, precisely to our *evidence*.

Husserl thus returns us to the modes of life of presence, of evident perception, which offers itself prior to any scientific operation, to that world always at hand and possessing an unmovable validity for our daily praxis. To establish a proper beginning, it is necessary to effect a *suspension of scientific validity or of the opinions of all sciences*: an *epoché* of all theories, to bracket them temporarily and suspend their use[6]; it would be a matter of choosing to *live first*, or even outside of any scientific consideration in a world that is therefore *pre-scientific*.

Through the *epoché*, we discover that, beyond any kind of *substruction*[7], we are already situated. The beginning discussed earlier has already occurred: it consists in the "*I already am*," in the "*I already exist*." The world in which we are situated is no mystery to us; rather, it is the world in its most obvious sense, the world with which we deal at every moment and to which our actions always refer.

But how is the world given to us? This question, the universality and scope of which cannot be evaded, examines how we live in the world, how it meets us, and what properly manifests itself to us of it. The world under reflection here is no longer the world of common opinion and unexamined habits, but the concrete world as lived in each moment of presence: the concrete world of life, of living perception, of living intentionality. It is precisely in perception that we find

---

6 See Husserl, E., *Ideas Pertaining to a Pure Phenomenology and to a Phenomenological Philosophy*. 1913, quot., pp.33-48.

7 See Id., *The Crisis of European Sciences and Transcendental Phenomenology: An Introduction to Phenomenological Philosophy*,quot., p. 38.

originally the evidence and our inherent relation to the world: the perceptual act paradoxically reveals the impossibility of a singular experience of the being-for-itself of an object that is not co-implicated in the worldly totality[8].

Hence, even when attending to a single object, the totality of the world is always co-implicated: the "I" and the world, in their intentional relation and reciprocal constitution, form the original phenomenological situation. Husserl leads us, therefore, to an awareness of the impossibility of conceiving ourselves independently of a certain relation to the world.

In *The Crisis of European Sciences*, his final work, the appeal to pre-predicative experience becomes properly a return to the life-world, the *Lebenswelt*. This remains a logical-transcendental problem, akin to the foundation of predicative thought in pre-predicative experience; yet here, the problem assumes two distinct yet interrelated directions. On one hand, the return to the *life-world* is a demand upon scientific thought, which is in crisis due to having lost all connection with its origins and intrinsic purpose, both rooted in the life-world. Only thus can science, abstracted into objectivity and pure *technization*, be restored to the level of philosophy – reclaimed in full awareness of its meaning[9]. On the other hand, the return to the life-world signifies the full thematization of the 'subjective' or 'egological' element, the *constitutive a priori* that anonymously underlies the self-evident structures of the life-world. Rendering this 'subjective' or 'egological'[10]

---

8 See, in this regard, the theory of Gestalt psychology mentioned above.

9 See Husserl, E., *The Crisis of European Sciences and Transcendental Phenomenology: An Introduction to Phenomenological Philosophy*, quot., pp.46-48.

10 The meaning of the term "subjective" or 'egological' is explained by Husserl in *The Crisis of the European Sciences and Transcendental Phenomenology*: "The epoché and the pure gaze aiming at the functioning egological pole, and therefore at the concrete totality of life and its intermediate and final formations, reveal nothing human eo ipso. [...] The ego that I draw upon in the

element explicit – hidden in the naïvely objectivistic attitude of common sense and abstractly scientific thought – is the task of a new science, more precisely a transcendental phenomenology of the life-world.

It is evident that the satisfaction of this second, properly transcendental demand is connected to the first, the requirement to restore the sciences to the sense of their origin and intrinsic purpose. The failure of science, which in its Renaissance beginnings was animated by faith in a "universal philosophy," stems from its progressive estrangement from philosophical problems. Mathematical natural science, originating in Galilean physics, lost its meaning through its "*technization.*" From this abyss of meaninglessness, science may ascend again by retracing its theoretical path, reconnecting with its origins. The *historical sense of the original foundation* can only be acquired and discovered by defining, in its structure and sense, human original experience: the experience of the life-world, the foundational ground from which every strictly scientific inquiry, every process of abstraction and idealization, begins.

The task of phenomenology is thus to investigate scientifically how the life-world functions as a foundation, *the way in which its prelogical validities are founded with respect to logical-theoretical truths.* The *objective and true world* of science is a theoretical-logical construction that presupposes and simultaneously refers back to the *Lebenswelt*; it is a construction upon, or – as Husserl says – a "substruction", something not immediately perceptible, yet founded upon the perceptible. From logical-objective mathematical evidence, from the evidence of the natural sciences, the path leads back

epoché, and which is equivalent to the ego that could be reached through a renewal and critical rectification of Descartes' epoché, is referred to as the 'I' only by mistake." Die Krisis der europäischen Wissenschaften und die transzendentale Phänomenologie, ed. Walter Biemel, Husserliana 6 (The Hague: Martinus Nijhoff, 1954), 187–8 (my translation).

to the original evidence, in which the life-world is constantly *already given*.

The concrete life-world is the ground upon which the *scientifically true world rests*, yet it simultaneously encompasses that world. The a priori of the pure life-world precedes the logical-objective a priori. Thus, a first *epoché*, through which one is removed from the validity of all objective sciences in order to isolate in its prescientific purity the general structures of the life-world, is necessarily followed by a second, *total* epoché, which brackets all natural life oriented toward *worldly* realities. Through the transcendental reduction, a radical transformation occurs (for us), leading to the discovery of the constitutive function of the I (*Ego Cogito*) within the life-world. In the natural attitude, oriented directly toward objects of the world, this function remained hidden: the ego, functioning anonymously, was absorbed within the horizon of the world. In the new perspective produced by the second reduction, the direction of attention is transformed. Thus arises a new type of science, distinct from other 'objective' sciences arising within the world: a science of the universal how of the already-given world, a science investing subjectivity for which the world is already given. Through this investigation, we can uncover the universal correlation, absolutely complete and autonomous, of world and consciousness-of-world.

Consequently, for Husserl, phenomenology redeems the failure of transcendental philosophy and the failure of psychology, for it unifies them in a sense; phenomenology resolves the fundamental paradox of the necessary difference and necessary identity of the transcendental and psychological ego. Hence, at the conclusion of his lecture entitled *Psychology in the Crisis of European Sciences*, he asserts: "It is necessary, therefore, to show how, in transcendental speculation, an obscure and unconscious intuition of the depth of the subject's life seeks expression, which is impossible to explicate in the objective attitude, and how this attempt must necessarily fail for lack of an analytic method; on the other hand, how psychology could not attain its proper theme, the very essence of the psychic sphere, as long as it remained within the objectivistic attitude and methodological prejudice characteristic of the natural sciences. The curious interrelation of the definitive failure of transcendental philosophy and the failure of psychology must first be demonstrated – not because they were connected, and thus destined to share the same fate, but precisely because they were separate. With this notion, the task is posed to free psychology from the nets of naturalistic objectivism and to initiate a transcendental philosophy as an analytic method of concrete investigation and explicit articulation of that subjectivity which a reformed psychology must first elaborate."[11]

Yet this very phenomenological reduction risks precisely a splitting of the world, forming a new immanent thought that, while offering the advantage of not privileging internal experience exclusively within the descriptive psychology of the mind, constantly threatens to revive a Platonic dualism and to give rise to the 'third-man' argument, whereby the Cogito is exposed to infinite regress – from form to content as in transcendental analysis, or from subject to object as in the reflective tradition. In Husserl, this difficulty manifests in the duality of eidetic and phenomenological reductions. Through the epoché, we divide ourselves into the one who lives and the one who observes our life. "Maintaining this *methodological schizophrenia*," Fink rightly observes, "requires great effort on the part of the thinker, particularly if it is to be more than a brief preliminary reflection, but rather a habitual research attitude."[12]

At the conclusion of the phenomenological-genetic analysis, hidden intentionality is unveiled, freed from anonymity, and

---

11 Husserl, E., *Manoscritto K III*, 1, it. transl. by E. Filippini, Bompiani, Milan 1960, pp. 107-108 (my translation).

12 Fink, E., *Les concepts opératoires dans la phénoménologie de Husserl*, La Haye 1959, p. 222 (my translation).

we understand ourselves as transcendental subjectivity: not as one facing the world, as in simple psychological reflection, but as one bearing within itself and capable of performing all operations to which the world owes its being.

The task of phenomenology is thus to liberate us from all sedimentation. Yet the reason for such sedimentation is difficult to discern. Husserl asserts that the authentic task of transcendental philosophy is not the absurd attempt to pass from some fantastical immanence to an equally fantastical transcendence, to locate an entity in itself, but rather the 'discovery of functioning intentionality and the analysis of its functioning.' Nevertheless, despite these speculative efforts, in Husserl's work the status of the transcendental ego seems not entirely persuasive. When he treats the transcendental relation, he continually slips toward analyses of the object as such, while simultaneously affirming the impossibility of legitimately executing them without constant reference to the subject.

This suggests that Husserl's Ego Cogito, the ego-horizon – the originating ground within which the manifestative encounter of self and world occurs—may not have fully overcome a Platonizing, consciential idealism: the founder of the phenomenological method himself appears unable to escape a form of transcendental idealism. This is because he endowed consciousness not merely with the capacity to confer meaning upon reality, but also with the power to constitute it. Moreover, as Jeanson observes, Husserl "forgot to reopen the brackets with which he had initially suspended the world, when it was necessary to liberate it from pseudo-objective interpretations (…) In asserting the principle of intentionality, Husserl marks the true origin of the phenomenological method, yet he can be reproached for betraying his own principle, disregarding the essential character of intentionality. In fact, in order to reveal the intentional movement characteristic of consciousness (every consciousness being consciousness of something), Husserl provisionally bracketed the world – yet he was subsequently seduced by the temptation to reconstruct it outside any factual situation, without reopening the brackets; a sign of transcendental idealism, or equivalently, of realism of essences (essence becoming an *idea* in the Platonic sense of *archetype*). The real world is thus reduced to the unreality of the perceived world; consciousness becomes the sole reality, and the objects of the world depend upon it not merely in the meaning it confers, but in their very being – over which its revealing power transforms into a *constitutive* faculty. Husserl insists on prolonging the phenomenological reduction until the free horizon of phenomena, considered in their transcendental purity, is discovered – thus arriving at the domain of phenomenology in the specific sense intended. From this perspective, he declares, a *transcendental-phenomenological* idealism is constituted (…). As Merleau-Ponty notes, however, *it would be necessary to become the transcendental subject in order to legitimately affirm this* – a condition evidently impossible[13].

Here, then, lies the profound problem intuited by Husserl yet unresolved. While we investigate the how of the world in the reflective attitude described, and uncover its relation to our spiritual and corporeal activities, are we not simultaneously within that world, part of it, objects among its objects? We are confronted with a paradox: we are divided, both subjects and objects, outside of a totality of which we are also a part. This division is internalized in our very corporeality: there is a body we move and direct in kinaesthesia, a body that acts in the world according to intentionalities emanating from our will (our living body, or *Leib*); and there is a body that belongs not so much to us as we to it, one upon which we depend and are rendered objects, a body living according to its own intentionalities, unknown to us, a body that ages, suffers, and conditions us just as external things of the so-called natural

13 Jeanson, F., *La fenomenologia*, Sugar, Milano 1962, p.138 (my translation)

surrounding world condition us (body-object that is part of the world and its circumstances, referred to by Husserl as *Körper*).

Thus, reflection upon the life-world – the world of living presence – already *presupposes that world as operative within us (embodied beings)*. In investigating the world as lived in evidence – the life-world – I myself become a problem: I am, insofar as I have this world, that which is interrogated.

From this problematic sphere – the *shadow* of Husserl – arises for Merleau-Ponty the need for a questioning reflection that surpasses the previous one (to which he is nonetheless deeply indebted), a reflection that interrogates not the generic life-world but myself as life in which the world manifests, my being-I as the total operative will directed toward the world according to the style of worldly certainty and quotidian validity. I must therefore ask not only *how the world is given to me, but how I have the world, and in what manner I possess this having*. Attention is now directed to the field of my intentional activities, insofar as they have a constant and internal reference to the world. Only at this point does our inquiry seem fully self-coherent, because only now do we return entirely and adequately to ourselves in the real situation in which we have been from the outset: that of philosophical reflection. The radicality of this reflection consists in suspending the results of the perceptual operation under investigation, not taking for granted as evident that the object exists and is perceived, but suspending this common validity to turn toward the acts through which precisely such validity is constituted and manifests itself.

The perceptual relation between myself and the object, as Merleau-Ponty will argue following Husserl, is not a relation between two "entities" that first live independently and then connect; the object has no reality for me outside my perception, nor am I an "entity" with reality independent of living perception of the world. The initial "ambiguity" of perception and any intentional act consists precisely in this: I and the world are inextricably intertwined, even if in common opinion we count as separate, as fully realized and independent realities; perception is simultaneously an act of my life, yet contains an internal reference to something *other than me*: it presents the object here, in my representation, yet it exists there, outside my representation.

Thus, the knot of the problem passes to Merleau-Ponty. It will be the task of the French philosopher to resolve the question left open by Husserl, taking up the *unthought-of element that opens out on something else*[14].

§ 3. *Merleau-Ponty's portrait*

From Husserl's phenomenology, discussed in the preceding paragraph, Merleau-Ponty takes up certain themes which, in accordance with their nature and the manner of their deployment, allow for an initial characterization of his thought. The attempt to establish a new philosophical perspective, beyond traditional frameworks, engages the French philosopher in a task of radical rethinking of the concepts upon which philosophical thought has been constituted and developed over its millennial history. One thus witnesses a continuous excavation, both in depth and in breadth, in search of new vantage points capable of liberating philosophical categories from the opposition and isolation to which the most entrenched canons of Western philosophical reflection seem to consign them.

Moreover, as Robinet observes,

> the Hegelian and phenomenological matrix of his reflection is too evident and too candidly acknowledged for one not to have described this situation, from the outset, in terms of the three H's: Hegel, Husserl, Heidegger. Yet, even this simple observation gives rise to an initial problem. It is not so much one of these authors who shapes his thought, but rather what was most

14 Merleau-Ponty, M., *Signs,* Northwestern University Press, Evanston, Ill.1964, p. 160.

> inwardly fertile in their philosophical endeavour. For his references reveal a preference for the Hegel of an open, unsystematized dialectic; for the late Husserl rather than the early one; and for a Heidegger who would not have renounced the conjunction of Being and existence. He takes them up as his own thought awaits them, and from this he draws what he considers the only living method in the history of philosophy: to grasp the path of an author's thought, the genesis beneath the structures, the becoming that traverses the essences.[15]

As is evident from his first major work, *The Structure of Behaviors*, the French philosopher seeks first and foremost to confront the presuppositions of the experimental sciences, in the attempt to discern a new task for philosophy and for thought as such. Thus, the problem of the relation between science and philosophy arises here in all its compelling force.

The fundamental critique that Merleau-Ponty directs against psychologistic sciences (already discussed in §1 of this chapter) is clearly articulated in the third chapter of the aforementioned work:

> It is not seen that, from the moment behavior is considered "in its unity" and in its human meaning, one is no longer dealing with a material reality nor, moreover, with a mental reality, but with a significative whole or a structure which properly belongs neither to the external world nor to internal life. It is realism in general which must be called into question.[16]

Thus, according to Merleau-Ponty, a proper notion of behavior can attain its philosophical status only by abandoning causal or mechanical thought in favour of a dialectical one. On a strictly epistemological level, he acknowledges the great progress achieved by *Gestaltpsychologie* in comparison with behaviourist theories. Against Watson's mechanistic approach,

---

15 Robinet, A., *Che cosa ha veramente detto Merleau-Ponty*, it. transl., Ubaldini Editore, Roma 1973, p. 49 (my translation).

16 Merleau-Ponty, M., *The Structure of Behavior*. Pittsburgh, Duquesne University Press, 2015, p. 182.

in which human psychological experience is reduced to a mere object of experimental observation, Gestalt theory has asserted the necessity of adopting the point of view of the human totality.

Yet this is not sufficient to make the position of Köhler and Koffka coincide with that of Merleau-Ponty. For, according to the French philosopher, even *Gestaltpsychologie*, by relegating the perceptual moment to a secondary place with respect to the construction of wholes, ultimately arrives at a form of objectivism of a naturalistic type. It is precisely the need to bring into focus the primacy of the perceptual moment understood as the unitary moment of the way in which the human being feels and lives – that opens for Merleau-Ponty the path toward his second major work, published in 1945, *Phenomenology of Perception*[17].

As several scholars of Merleau-Ponty's thought have pointed out[18], there exists a profound thematic continuity between *The Structure of Behavior* and *Phenomenology of Perception*, despite the shifting of perspectives and methodologies of inquiry. The speculative endeavour of the first work unfolds predominantly on the terrain of scientific description, with the intent of highlighting the philosophical limits underlying certain methodological presuppositions of the human sciences.

In *Phenomenology of Perception*, the philosopher situates himself at a level anterior to scientific knowledge, within that original perceptual experience which underlies the subsequent theoretical elaborations of science. The experience to which this work refers is that naïve and immediate experience amply thematized in Husserl's later writings. Merleau-Ponty speaks of it as of a world anterior to reflective knowledge – a world

---

17 Id., *Phenomenology of Perception,* transl. by Colin Smith, Routledge, London and New York 1962.

18 See Bonomi, A., *Esistenza e struttura. Saggio su Merleau-Ponty*, Il Saggiatore, Milan 1967, in particular pp. 23-85; and Tilliette, X., *Merleau-Ponty ou la mesure de l'homme*, Seghers, Paris 1970, in particular pp. 19-49.

which reflective thought describes, all the while knowing that it does so only by means of an extreme simplification.

The primordial experience of perception, says the philosopher, stands in relation to its scientific representation as a landscape stands to the geographical map that depicts it. From a certain point of view, then, as De Waelhens observes, "the perspective of The Structure of Behavior remains subordinate to that of the Phenomenology of Perception, just as the scientist's experience remains subordinate, in its origin, to the everyday experience it seeks to explain and without which it would not subsist."[19] Yet, from another point of view, it is also true that we ordinarily move within the network of conventions established by society, and the return to the immediacy of perceptual experience is realized only at the end of a reversal – the methodological presupposition of which, for Merleau-Ponty, is the phenomenological reduction.

In the years that followed, the philosopher published no systematic work comparable to The Structure of Behavior or the Phenomenology of Perception. His inquiry instead moved toward particular deepening and re-elaboration of insights already present in those two major works. This does not mean, however, that Merleau-Ponty's philosophy after 1945 represents a mere repetition of positions already acquired. Following Husserl's own mode of proceeding, he continually returns to the fundamental nuclei of his thought, seeking to discern within them ever-new implications and possible developments.

From this second phase of research arise the essays later collected in *Sense and Non-Sense*[20] and in *Signs*[21],

19 De Wahelens, A., *Una filosofia dell'ambiguità*, Preface to *La struttura del comportamento*, Bompiani, Milan 1963, pp. 17-18 (my translation).

20 Merleau-Ponty, M., *Sense and Non-Sense*, Northwestern University Press, Evanston, Ill. 1992.

21 Id., *Signs*, quot.

the predominantly political work *Humanism and Terror*[22], the essay *Adventures of the Dialectic*[23], and the inaugural lecture delivered at the Collège de France on 15 January 1953, entitled *In Praise of Philosophy and Other Essays*[24]. A further moment of Merleau-Ponty's reflection – abruptly interrupted by his sudden death – is testified by the pages of the unfinished manuscript *The Visible and the Invisible*[25]. In this work, which critics unanimously interpret as the document of the metaphysical turn in Merleau-Ponty's thought, the philosopher does not renounce his previous positions; rather, he reconsiders them in confrontation with the problem of ontological foundation. The question of Being now becomes the horizon against which Merleau-Ponty's rethinking unfolds.

According to Nepi, what is at stake is a "phenomenological ontology whose immediate term of comparison is not to be found in the categories of classical metaphysics, but in Husserl's attempt to overcome the oppositions upon which that metaphysics – from Greek speculation to the great philosophical systems of the nineteenth century – was constituted."[26] The concept of being as objectivity, set in opposition to thought as subjectivity, represents for Merleau-Ponty, as for the entire development of Western thought, an intrinsic limit. The new ontological perspective he seeks to open thus situates itself beyond this opposition, aiming to ground a concept of Being that simultaneously embraces both the dimension of subjectivity and that of objectivity. Moreover, in developing the problem of the relation between

22 Id., *Humanism and Terror: An Essay on the Communist Problem*, Beacon Press, Boston 1947.

23 Id., *Adventures of the Dialectic*. Northwestern University Press, Evanston, Ill. 1973.

24 Id., *In Praise of Philosophy and Other Essays*, quot., 1988.

25 Id., *The Visible and the Invisible*, transl. by A. Lingis, edited by C. Lefort, Northwestern University Press, Evanston, Ill. 1968.

26 Nepi, P., *Merleau-Ponty tra il visibile e l'invisibile*, Studium, Rome 1984, p.18 (my translation).

metaphysical reflection and painting, he intimates yet another critique of classical metaphysics – one which, in his view, remains overly conditioned by the instruments of rational knowledge. All of this finds testimony in his essay *Eye and Mind*[27], published in the first issue of Art de France in January 1961. It is the last work Merleau-Ponty completed, written contemporaneously with the manuscript of *The Visible and the Invisible*.

To the works already mentioned we should add, for a complete reconstruction of Merleau-Ponty's intellectual itinerary, several other writings published posthumously – such as *The Prose of the World*[28], his university lectures, for which abundant critical documentation exists, as well as his various conferences and public debates. All these sources prove indispensable for any comprehensive interpretation of Merleau-Ponty's thought.

In the following pages, to complete our excursus along the Merleau-Pontian path, we shall analyse the decisive moments in the development of the French philosopher's thought—moments that presuppose continual reference to the philosophical orientation which most profoundly inspires him, namely, phenomenology. Of this tradition, and of its founder Edmund Husserl, Merleau-Ponty takes up and deepens certain perspectives: above all, the rigor of the method. For—as the philosopher himself writes – "It is the search for a philosophy which shall be a 'rigorous science', but it also offers an account of space, time and the world as we 'live' them"[29]. But Husserl's philosophy also articulated a critical investigation into certain specific perspectives, among which the concept of the *life-world* assumes a fundamental position. This theme,

27 Merleau-Ponty, M., *Eye and Mind*, in J. E. Edie (Ed.), *The Primacy of Perception*, Northwestern UP Evanston, Evanston, 1964, pp. 159-190.

28 Id. *The Prose of the World*, Northwestern University Press, Evanston, Ill. 1991.

29 Id., *Phenomenology of Perception*, quot., p. VII.

present throughout much of Husserl's reflection, acquires in Merleau-Ponty's thought – particularly in his early works – a primarily psychological significance. His first major works, *The Structure of Behavior* and, above all, *Phenomenology of Perception*, operate within the endeavour to configure a renewal in the conception of the logical-psychological structures of human experience, and this in close relation to the demands intrinsic to phenomenological inquiry.

Within this framework, the reference to Husserl's notion of the *life-world* signals an epistemological imperative. Yet in Merleau-Ponty, the category of the *life-world* already carries an undeniable ontological import. Indeed, in the methodological preface opening *Phenomenology of Perception*, introduced by the question "*What is phenomenology*?". Merleau-Ponty takes up the concept of the *life-world*, particularly emphasizing the necessity of transcending the classical distinction between logical essence and actual existence. From this it becomes evident that, in Merleau-Ponty, the ontological question is to be understood as arising precisely from the overcoming of the distinction between logical activity and real existence. "Phenomenology," the philosopher declares, "is the study of essences; and according to it, all problems amount to finding definitions of essences: the essence of perception, or the essence of consciousness, for example"[30].

But essences, as Merleau-Ponty immediately specifies, are not mere data of thought, nor abstract logical constructions. "*Phenomenology,*" he adds, "*is also a philosophy which puts essences back into existence and does not expect to arrive at an understanding of man and the world from any starting point other than that of their 'facticity*'"[31]. Merleau-Ponty's ontology thus aims to reach, while emphasizing their existential dimension, the presuppositions of phenomenological inquiry.

30 Ibid.
31 Ibid.

The world, and in particular the life-world, is for phenomenology neither the product of the logical categories of the natural sciences nor the Kantian noumenon opposed to the phenomenon. It is the original positivity that constitutes the horizon from which every perspective receives both its guarantee of objectivity and its meaning.

Phenomenology, therefore, in Merleau-Ponty's view, already contains in Husserl an indispensable existential dimension. It is, then, a philosophy of essences – provided that essence is not understood in an abstractly logical but in an existential sense. From the phenomenological standpoint, essence and existence are not two opposed categories. The overcoming of the distinction between essence and existence thus corresponds to the overcoming, which phenomenology has sought since its very inception, of the classical distinction between subject and object. Merleau-Ponty's interpretation of Husserl thus moves in a distinctly "existential" direction – though not in the sense of existentialism.

Here already the theoretical significance underlying Merleau-Ponty's reconsideration of Husserl's life-world becomes evident. It is true that the theme of the reduction to the life-world is conceived by Merleau-Ponty, particularly in the initial phase of his research, as instrumental to the elaboration of new logical and psychological structures of the human being. Yet the existential positivity of the life-world, as the primordial horizon from which to reconceive the traditional opposition between subject and object, becomes the presupposition for a broadening of perspective from the logical to the ontological plane. The concrete existentiality of the life-world, regained beyond the abstract formalization of scientific thought and the schematic oppositions of past philosophies, provides the premise for a renewed comprehension of Being in general (the life-world assumes, in Merleau-Ponty, a preeminent role compared to the initial Husserlian perspective of eidetic reduction, within which, however, it's possible to identified the very premises of that *Lebenswelt* which would later, especially in *The Crisis of the European Sciences*, become the object of extensive thematic elaboration)[32].

It should therefore be noted how this existential dimension of phenomenology deepens in the development of Merleau-Ponty's work. The perspective emerging from *Phenomenology of Perception* and found again in the collection *Sense and Non-Sense*, brings us into contact with a concept of metaphysics that, in certain respects, is more indeterminate than that which appears in his later writings, particularly in *The Visible and the Invisible*.

In the first phase of Merleau-Ponty's reflection, the concept of metaphysics is defined in relation to science. The model of science to which he refers is provided above all by *Gestalt psychology* and by the emerging structural epistemology, upon which the modern human sciences – psychology, sociology, linguistics, and others – would later be founded. But what is the relationship established between these and metaphysical knowledge? As Merleau-Ponty writes, "*As presently oriented, the sciences of man are metaphysical or transnatural in that they cause us to rediscover, along with structure and the understanding of structure, a dimension of being and a type of knowledge which man forgets in his natural attitude*"[33].

The interpretation of the concept of structure – in which, according to the French philosopher, science and metaphysics converge – acquires, in this initial period of his thought, the meaning of a return to the original dimension of existence. Confronted with the laws and conventions accredited by society, and which common sense now regards as natural, the concept of structure performs a kind of bracketing. In later and more elaborate developments, the concept of structure would undergo a profound theoretical reworking, eventually

32 See Zecchi, S., *La fenomenologia dopo Husserl nella cultura contemporanea*, voll. 2, *La Nuova Italia*, Florence 1978, in part. vol. 1, pp. 48-56.

33 Merleau-Ponty, M., *Sense and Non-Sense*, quot., p. 92.

taking on the form of a logical-mathematical model. Yet in its first formulation, structure expresses a regression toward the elemental, the primitive – toward that which, beyond the conventional social and cultural distinctions, unites the original roots of being and of life. In a certain sense, therefore, the concept of structure appears at this moment in Merleau-Ponty to align itself with the same direction as the phenomenological reduction – a path, as has been seen, intentionally open to the recovery of the ontological dimension of existence.

A little more than a decade later, Merleau-Ponty would return to the ontological problem, yet the general horizon of his thought had by then undergone a kind of radicalization. The ontological turn of the later Merleau-Ponty situates itself within a precise trajectory of contemporary thought. In the background of *The Visible and the Invisible* one perceives the influence of Sartre, of Heidegger, of Bergson's intuitionism, and of the metaphysical themes of existentialism.

Merleau-Ponty's ontology thus presents itself with unmistakable traits in relation to the ontological thinking that Western philosophy has articulated throughout its history. In Merleau-Ponty – as in Heidegger and in other philosophers of our time – together with a vivid sense of the problem of Being there is also a refusal to offer an intellectualistic solution, that is, one founded on the presupposition of an *adaequatio* between being and thought. This does not mean, of course, falling into the domain of art or literature; rather, the ontology that proceeds from such a perspective is precisely a "semantic ontology"[34], woven of signs and symbols, because the philosopher's task is defined – without ever being exhausted – within the complex exercise of philosophical interpretation, or hermeneutics.

To attain this kind of ontology, Merleau-Ponty insists, it is necessary to overcome the various presuppositions upon

34 See Nepi, P., *Merleau-Ponty tra il visibile e l'invisibile*, quot., pp. 132-144.

which traditional metaphysics has been constructed. First among them is the presupposition of attaining the absoluteness of truth, which, from Merleau-Ponty's standpoint, represents the destruction of critical thought and hence of philosophical inquiry: as he writes, "*Metaphysical and moral consciousness dies upon contact with the absolute*"[35].

Philosophy's claim to reach truth is, for Merleau-Ponty, an unjustified prejudice. The proof of this, he suggests, lies in the religious outcomes to which all so-called absolute thinkers have inevitably been led. The very notion of the absolute thus constitutes, for Merleau-Ponty, the death of philosophy; and in its place the French philosopher substitutes the concept of contingency. Indeed, "*The contingency of all that exists and all that has value is not a little truth for which we have somehow or other to make room in some nook or cranny of the system: it is the condition of a metaphysical view of the world*"[36].

Yet the contingency of which Merleau-Ponty speaks is not, as in Sartre, the empty dialectic of being and nothingness. Rather, it is the awareness that Being, instead of expressing the fullness of the absolute, constitutes the very horizon of our thoughts and actions – so that our relation to it should not take the form of appropriation, but of search. It is not by chance, then, that philosophical inquiry expresses itself, for Merleau-Ponty, in the very form of questioning.

> The philosophy placed in books has ceased to challenge men. What is unusual and almost insupportable in it is hidden in the respectable life of the great philosophical systems. In order to understand the total function of the philosopher, we must remember that even the philosophical writers whom we read and who we are have never ceased to recognize as their patron a man who never wrote, who never taught, at least in any official chair,

35 Merleau-Ponty, M., *Sense and Non-Sense*, quot., p. 95.
36 Therein, p. 96.

> who talked with anyone he met on the street, and who had certain difficulties with public opinion and with the public powers. We must remember Socrates.[37]

Metaphysics, then, is a questioning that does not conceive of an answer capable of cancelling it, but only of resolute actions that carry it further. The ontological perspective that descends from such a conception of philosophy is therefore more an attempt than a result, and it takes shape – within that original re-reading of Husserl which characterizes Merleau-Ponty's phenomenological method – in constant dialogue with Husserl's concept of intentionality. The latter, which in Husserl assumes both an eidetic and a genetic character, is interpreted by Merleau-Ponty primarily in its genetic sense. It expresses, then, not so much the relation between the activity of thought and the world of essences, as rather the movement of thought toward a reality that is itself movement.

For this reason, the human being – who alone is the 'metaphysical animal' – is defined by Merleau-Ponty, "as opposed to the pebble which is what it is, is defined as a place of unrest (*Unruhe*), a constant effort to get back to himself, and consequently by his refusal to limit himself to one or another of his determinations"[38].

Philosophical knowledge, rather than a completed and absolute knowing, is therefore above all a non-knowing – where the negative formula, for Merleau-Ponty as for Socrates, does not imply a sceptical premise but rather expresses the transcendence of truth in relation to every determinate knowledge. The logic associated with semantic ontology is thus not a logic of the concept but a logic of the sign. For the concept exhausts reality within its rational contents, whereas the sign indicates a tendency, a direction of inquiry, and may

37 Id., *In praise of Philosophy*, quot., p. 34.
38 Id., *Sense and Non-Sense*, quot., p.66

require the surpassing – though not the abolition – of the rational structures of knowledge.

Sign and meaning are precisely the two terms that distinguish semantic ontology from intellectualistic ontology. Merleau-Ponty prefers to speak of the sign rather than of meaning. Meaning places in the foreground the activity of the constituting subject, whereas the sign, though it also implies a reference to the subject, begins to define itself within things themselves, expressing their own law of coherence. Consequently, semantic ontology situates itself beyond both idealism and realism – the two philosophical modes through which traditional metaphysics has been configured.

> There are – Merleau-Ponty writes – two classical views: one treats man as the result of the physical, physiological, and sociological influences which shape him from the outside and make him one thing among many; the other consists of recognizing an a-cosmic freedom in him, insofar as he is spirit and represents to himself the very causes which supposedly act upon him. On the one hand, man is a part of the world; on the other, he is the constituting consciousness of the world. Neither view is satisfactory.[39]

To the first perspective would belong the various forms of empiricism and materialism, including the Aristotelian-Thomistic metaphysical tradition whose principal defect, for Merleau-Ponty, lies in its naturalism; to the second would belong the different forms of idealistic philosophy, to which Sartre's existentialism would in the end be reduced. Conversely, within a philosophy of the sign, semantic ontology represents a standpoint that transcends the unilateral reductions to either the sphere of the objective or that of the subjective. It is only in this precise sense that it may be qualified as a "philosophy of ambiguity."[40] This expression designates precisely the co-

---

39 Therein, pp. 71-72.

40 De Wahelens, A., *Una filosofia dell'ambiguità*, Preface to *La struttura del comportamento*, quot., p. 23 (my translation).

presence of a subjective and an objective moment within every human activity, beginning with cognition itself.

"The subject," observes Enzo Paci in his Introduction to *Sense and Non-Sense*, "is not an absolute observer, detached from the world; the object is not a transcendent reality, detached from the way in which human beings perceive it"[41]. This correspondence between the subjective and the objective, however, does not occur in the traditional terms of *adaequatio*. This is because the subject is always a corporeal subject and thus situated within a fabric of worldly and historical relations. In the same sense, the object can never be defined apart from the perceptual activity of consciousness. If intentionality expresses the objective rootedness of cognitive activity, the sign expresses, rather, the appearing of an object within the perceptual horizon of consciousness.

Semantic ontology thus reaffirms itself as one of the keystones of Merleau-Ponty's philosophy: and the being it defines is simultaneously the being of the perceiving subject, the being of the perceived object, and, more originally still, that of their mutual relation.

Merleau-Ponty's interpreters are largely agreed in recognizing, within his later writings, a decisive ontological orientation. Yet it is precisely on this point that divergence immediately arises. Is it a rupture with his earlier phenomenological position? A continuity of the same line of inquiry? A mystical withdrawal? A scholar, named R. C. Kwant[42], has characterized this development in Merleau-Ponty's thought as the transition *from phenomenology to ontology*. It would thus be neither a break, nor a simple continuity, nor a regression, but rather the explicit unfolding

41 The reference can be found on p. 11 of Introduction by Enzo Paci to Merleau-Ponty's work Sense and Non-sense, quot.

42 Kwant, R. C., *From phenomenology to metaphysics. An inquiry into the last period of Merleau-Ponty's philosophical life*, Duquesne University Press, Pittsburgh 1966, in part. pp. 228-231

of those nuclei which, in the earlier works, had been only intimated.

That the ontological problem is, in fact, always present in Merleau-Ponty is attested by the frequent use of ontological categories already in his first works. In *Phenomenology of Perception*, for example, one often encounters the concept of *being-in-the-world* (*être-au-monde*), which Merleau-Ponty examines primarily in its logical and psychological dimensions, without thereby neglecting its ontological significance. Yet, in his later writings, the concept of *being-in-the-world* tends to be abandoned and replaced by other notions – such as *being-of, the flesh*[43] *of the world*, and so forth.

In his early works, then, ontological categories are implicitly embedded within the phenomenological procedure aimed at describing the structures of behavior and perception. In one way or another, Merleau-Ponty was always aware that his perspective carried an undeniable ontological resonance. How else could we explain his certainty that philosophy must overcome the traditional opposition between subject and object, between thought and being, between the *I* and the world?

---

43 On the concept of flesh see in particular: Alloa E., *La chair comme diacritique incarné,* in "Chiasmi International", n. 11, 2009, pp. 249-262; Barbaras R., *Les trois sens de la chair. Sur une impasse de l'ontologie de Merleau-Ponty,* in "Chiasmi International", n. 10, 2008, pp. 19-33; Kristensen S., *Corps et symbolisation. La phylosophie du dernier Merleau-Ponty et la question d'une épistémologie de la chair,* in "Chiasmi International", n. 11, 2009, pp. 321-337; Kristensen S., *L'incoscient machinique et l'idée d'une ontologie politique de la chair,* in "Chiasmi International", n. 16, 2014, pp. 127-147; De Saint Aubert E., *Les "différenciations d'une seule et massive adhésion a l'être qui est la chair". Étude de la notion de polymorphisme chez Merleau-Ponty,* in "Chiasmi International", n. 11, 2009, pp. 185-199; Michalet J., *La chair comme "plissement du dehors". La lecture deleuzienne du dernier Merleau-Ponty,* in "Chiasmi International", n. 13, 2011, pp. 241-258.

Let us now attempt to clarify the ontological significance of Merleau-Ponty's concept of *being-in-the-world*, to see how it is later taken up and radicalized in his final works. It will emerge that Merleau-Ponty, while adopting this concept within the framework of the phenomenological perspective, assigns to it a meaning that ultimately cannot be reduced to Husserl's position. Husserlian intentionality posits the world as the referential pole of consciousness. The world, therefore, cannot be conceived without an ego, and the ego exists only insofar as it is inherent in the world. It is, of course, a *worldliness* purified – through the moment of reduction – of every naturalistic residue. In Merleau-Ponty, one perceives both the need to remain faithful to the rigor of Husserl's phenomenological method and the need to transcend the limits of a philosophy that, in its opposition to psychologism, sometimes risks falling into the formal abstractions of logicism. The task, in fact, is not merely to determine the perceptual and cognitive dimensions of our being-in-the-world, but also to uncover the existential implications of the human-world relation.

As already suggested at the end of the previous paragraph, it is precisely the existential motifs already present in Husserl's thought that Merleau-Ponty seeks to emphasize and deepen in their ontological scope. Nevertheless, the category of *being-in-the-world*, as noted earlier, is ultimately replaced in the final phase of Merleau-Ponty's reflection. What is preserved from it is its original intentionality – its aim of clarifying the problem of distinction and encounter between the poles of ego and world: a distinction that never becomes separation, and an encounter that never becomes identification.

Yet, once the ontological presuppositions of Merleau-Ponty's thought are made explicit, the category of *being-in-the-world* is replaced by another, judged more adequate to express the new standpoint. The new category, which in *The Visible and the Invisible* becomes the structural foundation of Merleau-Ponty's ontology, is that of *being-of* (*en-être*).

This ontological radicalization, compelling Merleau-Ponty to bring to light the original dimension of both self and world, reveals the limitations of the earlier category of *being-in-the-world*. In that notion, where the phenomenological perspective prevails, the question concerning the primordial structure of self and world remains unarticulated, while the relation between the reality of the self and that of the world is described in terms of existential analysis.

The later Merleau-Ponty is no longer content to take for granted, as he once did, the existence of man and the world; he wishes instead to pose, even on this level, the fundamental question of philosophy itself. The category of *being-of* does not merely express the need to deepen and radicalize the bond of inherence between self and world. Since it denotes relation and participation, it inevitably refers to that original ontological ground upon which both the objectivity of the world and the subjective experience of consciousness are constituted.

Thus Merleau-Ponty writes of this necessity to radicalize philosophy's very questions:

> The philosophical interrogation is therefore not the simple expectation of a signification that would come to fill it. "What is the world?" or, better, "what is Being?"— these questions become philosophical only if, by a sort of diplopia, at the same time that they aim at a state of things, they aim at themselves as questions — at the same time that they aim at the signification "being," they aim at the being of signification and the place of signification within Being. It is characteristic of the philosophical questioning that it return upon itself, that it ask itself also what to question is and what to respond is. Once this question to the second power is raised, it cannot be effaced. Henceforth nothing can continue to be as if there had never been any question.[44]

The later Merleau-Ponty, no less than the author of *Phenomenology of Perception*, thus continues to assign to philosophy the task of interrogating the pre-reflective world of perceptual experience. Yet, in order to grasp the relation

44 Merleau-Ponty, M., *The Visible and the Invisible*, Northwestern University Press, Evanston, Ill. 1968, pp. 119-120.

between this lived, pre-categorical world and the philosophical question that seeks to unveil it, the former categories no longer suffice. It is as though the philosophical question itself were subjected to a kind of transcendental reduction squared – a reduction carried to its own limit – beyond which, unless it places itself in question, it inevitably lapses into contradiction.

Within this vast and problematic terrain takes root and gains form the final phase of Merleau-Ponty's thought – the most fertile in profound intuitions and laden with future developments – a phase which death abruptly cut short, leaving it to posterity as a fertile legacy that still requires a huge amount of work.

As Sini, an acute interpreter of Merleau-Ponty, writes:

> Here, the aim is to reweave a texture of thoughts, a map of itineraries, without any pretence of offering the reader a definitive image or snapshot of what Merleau-Ponty might have been, or perhaps, in part and secretly, was. Rather, the intention is to suggest the meaning – for us – of a path through which Merleau-Ponty, perhaps finding himself and perhaps losing himself within it, reaches into the very contemporaneity of our own problems. In doing so, he draws us into his questioning and, with a kind of generosity, leaves to us the task of uttering the words of a fulfilment that is, fortunately, impossible.[45]

After this preliminary overview of Merleau-Ponty's overall thought process, let us now take a closer look at it, almost step by step, to identify and better understand its individual stages.

---

45 Sini, C., *Introduzione* to M. Merleau-Ponty, *La Prosa del mondo*, Editori Riuniti, Rome 1984, pp. 7-8 (my translation).

CHAPTER 2
RE-FLECTING THE BODY, UN-FOLDING THE WORLD

§ 1. *"Seek Experience at its Source"*

> Everyone must question himself carefully on this point, after making a clean sweep of everything which his past experience has taught him about the cause of his sensations and coming face to face with the sensations themselves. […] Put aside what you remember of your past experiences and what you are accustomed to say of the present ones. […] It would be to seek experience at its source, or rather above that decisive turn where, taking a bias in the direction of our utility, it becomes properly *human* experience.[1]

The necessity to reverse the habitual course of thought is not merely a demand felt by Bergson, but constitutes the speculative commitment of a thinker such as Merleau-Ponty, who assumes the task of overcoming both idealistic and realistic claims: "It is a question of recognizing consciousness itself as a project of the world, meant for a world which it neither embraces nor possesses, but towards which it is perpetually directed – and the world as this pre-objective individual whose imperious unity decrees what knowledge shall take as its goal"[2].

It is therefore necessary to demonstrate how empiricism and rationalism sever the two poles of a structural relation –

1 Bergson, H., *Matter and Memory*, transl. by N. M. Paul and W. Scott Palmer. London, Allen Unwin, 1962, p. 241.

2 Merleau-Ponty, M., *Phenomenology of Perception*, quot., p. XX.

the former by resolving subjectivity into a concatenation of external occurrences, of natural events, the latter by confining the truth of the world within the interiority of consciousness, in order to definitively overcome the impasse of a thought determined either objectively or subjectively.

According to Merleau-Ponty, the problem of empiricism lies in explaining the complex through the simple, thereby depriving perception of the faculty to "usher knowledge" and fragmenting it into a myriad of sensations: the sensation-atom would thus constitute the simplest element, an ineffable and irreducible impression.

> All the difficulties of realism arise precisely from having tried to convert this original relation into a causal action and to integrate perception into nature. As soon as the presence or the presentation of a "thing" to consciousness – instead of remaining an ideal relation, as in naive experiences – is interpreted as a real operation of the thing on the body and on the perceiving subject, it becomes impossible to reconstitute the descriptive content of perception, the actual view of the world, as an effect.[3]

Empiricism is thus a form of thought that ignores itself and turns outward into the exteriority of things: in doing so, it conceals from us the horizons within which our life unfolds, both cultural and human, as well as natural. By reducing the perceived to a sum of physical properties, empiricism degrades the cultural world in which we have always been immersed to mere illusion, and at the same time transforms nature into a neutral assemblage of stimuli and qualities; consequently, it renders the primary ground of all experience unrecognizable. The analyses of empiricism take place within a preconstituted world, and their object is always embedded in a universe of 'already given things'.

3 Id., *The Structure of Behavior*, quot., p.193

> But not only does empiricism distort experience by making the cultural world an illusion, when in fact it is in it that our existence finds its sustenance. The natural world is also falsified, and for the same reasons. What we object to in empiricism is not its having taken this as its primary theme of analysis. For it is quite true that every cultural object refers back to a natural background against which it appears and which may, moreover, be confused and remote. Our perception senses how near is the canvas underneath the picture, or the crumbling cement under the building, or the tiring actor under the character. But the nature about which empiricism talks is a collection of stimuli and qualities.[4]

The empiricist, therefore, aspires to the absence of presuppositions, taking particular facts and observations as the origins of knowledge: such an attitude is certainly an effective weapon against convictions authoritatively imposed by rationalistic dogmas. Yet the rigorous application of the empirical method, which would entail limiting oneself to what is given in sensible experience, proves inadequate for the purposes of both science and philosophy; all the more so since it is well known that empiricism itself makes use of certain assumptions, such as recognizing the validity of principles transcending actual as well as possible experience, and the use of conceptual schemes. The idealist, on the other hand, in asserting the absoluteness of spirit, culminates in the form of an absolute consciousness, a transcendental ego, an all-encompassing mind:

> The radical originality of Cartesianism – Merleau-Ponty explains – is to situate itself within perception itself, not to analyze vision and touch as functions of our body, but "only the thought of seeing and touching." Beyond causal explanations which constitute the appearance of perception as an effect of nature, Descartes, in search of the internal structure, makes its meaning explicit and disengages the grounds which assure naive

4 Id., *Phenomenology of Perception*, quot., p. 60.

> consciousness that it is acceding to "things"; that, beyond the transitory appearances.[5]

If then, in empiricism consciousness was too impoverished and could do nothing but register a wholly exterior lawfulness, in rationalism it is too rich, defining itself as an abstract and constitutive power. Thus, in both cases – Merleau-Ponty emphasizes – the original act by virtue of which an object begins to exist for perception and becomes imbued with a certain meaning is lost from view:

> The perception of a distance or a size is not the same as the quantitative estimations by which science makes distance and size precise. All the sciences situate themselves in a "complete" and real world without realizing that perceptual experience is constituting with respect to this world. Thus, we find ourselves in the presence of a field of lived perception which is prior to number, measure, space and causality and which is nonetheless given only as a perspectival view of objects gifted with stable properties, a perspectival view of an objective world and an objective space. The problem of perception consists in trying to discover how the inter subjective world, the determinations of which science is gradually making precise, is grasped through this field. The antinomy [...] is based upon this ambiguous structure of perceptual experience. The thesis and the antithesis express the two aspects of it: it is true to say that my perception is always a flux of individual events and that what is radically contingent in the lived perspectivism of perception accounts for the realistic appearance. But it is also true to say that my perception accedes to things themselves, for these perspectives are articulated in a way which makes access to inter-individual significations possible; they "present" a world. Thus, there are things exactly in the sense in which 1 see them, in my history and outside it, and inseparable from this double relation. I perceive things directly without my body forming a screen between them and me; it is a phenomenon just as they are, a phenomenon (gifted, it is true, with an original structure) which precisely presents the body to

5 Id., *The structure of Behavior*, quot., p. 195.

> me as an intermediary between the world and myself although it is not as a matter of fact. I see with my eyes, which are not an ensemble of transparent or opaque tissues and organs, but the instruments of my looking. The retinal image, to the extent that I know it, is not yet produced by the light waves issuing from the object; but these two phenomena resemble and correspond to each other in a magical way across an interval which is not yet space. [...]
>
> The constituted world is confronted with the perceptual experience of the world and one either tries to engender perception from the world, as realism does, or else to see in it only a commencement of the science of the world, as critical thought does. To return to perception as to a type of original experience in which the real world is constituted in its specificity is to impose upon oneself an inversion of the natural movement of consciousness; on the other hand every question has not been eliminated: it is a question of understanding, without confusing it with a logical relation, the lived relation of the "profiles" to the "things" which they present, of the perspectives to the ideal significations which are intended through them [...] The natural "thing," the organism, the behavior of others and my own behavior exist only by their meaning; but this meaning which springs forth in them is not yet a Kantian object; the intentional life which constitutes them is not yet a representation; and the "comprehension" which gives access to them is not yet an intellection.[6]

It is therefore necessary to suspend both objectivist and subjectivist assumptions.

To return to things themselves is to return to that world which precedes knowledge, of which knowledge always speaks, and in relation to which every scientific schematization is an abstract and derivative sign-language [...] This move is absolutely distinct from the idealist return to consciousness, and the demand for a pure description excludes equally the procedure of analytical reflection on the one hand, and that of scientific explanation on the other.[7]

---

6 Therein, pp. 218-220 and 224.

7 Id., *Phenomenology of Perception*, quot., pp. IX and X.

But the problem is not merely to overcome, so to speak, the empiricist or rationalist claims; it is above all and primarily a matter of rejecting the 'certainty' of classical rationality, which for several centuries presented itself with the features and authority of a natural, necessary, and a priori structure. The power of its grasp was, in fact, from the very beginning, of unlimited breadth, because that reason was considered responsible for every cognitive possibility and every logical possibility: everything falling within the sphere of the specific, the individual, was thereby degraded in comparison with the austere rational structures of a central, exclusive, and invariant order, within which the play of all possibilities inherent in things, in nature, as well as in the movements of our thought, had already been codified and preconstituted. In this way, classical reason manifests itself as a *mathesis universalis*[8], as an ideal order already predetermined in things and thus enjoying a status of inexorable evidence (as, for example, was the fate of Euclidean geometry until the mid-nineteenth century). It is only through abstract, rigid, and unrelated rational schemes that the classical style of thought could erect its axiomatic system, within which to discipline every event and every form of experience.

Thus, reason exhibited itself as the absolute harmony of the cosmos and, at the same time, as the norm for the natural education of the mind. The human eye does not open upon the original darkness of the world, but upon the scene of a cosmos already illuminated by intellectual light.

As a result of this substantial transformation, of which the paradigms of classical reason have become the scaffolding of the entire reality, reason appears as a structure that is

---

8 The term *mathesis* is here understood in its etymological sense: from the Indo-European [...] root *mndh, men* derive the Greek word *mathesis* (knowledge, cognition) and the Latin word *mens* (mind).

twice *absolute* and twice *natural*. For it holds in itself and for itself, without reference to anything external, regulating all phenomena without expressing the natural lawfulness of thought, and at the same time the intrinsic lawfulness of nature and the universe.

If traditional rationality assumed the form we have outlined, this depended upon a strenuous and violent aspiration for an established and 'guaranteeing' order. That is, it depended on the circumstance that cognitive research was joined to a strategy aimed at 'uniting' both natural phenomena and intellectual, moral, and social conduct within a system of norms and paradigms for which no limits or exceptions were foreseen and thus assumed to be irrevocable and unconditional.

It was this strategy that configured human rationality as a kind of *nature*: on the one hand, as the nature of thought, or more precisely, as the "natural laws of thought", and on the other, as the objective structure of the world. In these terms, an essential feature of traditional rationality is its tendency to inscribe inquiry within the backdrop of an intellectual scene in which, for each question, the answer is already symmetrically provided.

At this point, detaching oneself from the truth-claims of classical rationality is, according to Merleau-Ponty, the first step to be taken toward a non-theoretical, non-impositive thought, and, as Sini says, "panoptic."[9] Even science, as we have already illustrated at the beginning of this work, no longer begins with self-evident premises.

According to Gargani, "Classical rationality entered into crisis when the central theme of its secret art was called into question – namely, the inspiration to represent the acts of intellectual conduct in the epic image of a sublime cosmic-

9 Sini, C., *Introduzione* a *La prosa del mondo*, Editori Riuniti, Rome 1984, p.15.

logical prelude, in which all the forms of what human beings can think and say are presupposed"[10].

The examination carried out so far should show that between things, facts, and processes on the one hand, and reason and language on the other, there exists neither the dichotomy of insurmountable opposition nor the natural and guaranteed correspondence that dogmatic rationalism presupposed. Yet the system of references by which thoughts and linguistic symbols are connected has too often been taken for granted, and this has led to serious consequences: indeed, a guaranteeing philosophical thought excludes the possibility of proceeding in a manner different from that of the *mente concipio* of Galilean memory.

> The impetus of reflection would overshoot its goal: it would transport us from a fixed and determinate world to a consciousness without a rift in it, whereas the perceived object is infused with secret life, and perception as a unity disintegrates and reforms ceaselessly. [...] In actual perception taken at its origin, before any word is uttered, the sign offered to sense and the signification are not even theoretically separable.[11]

With Merleau-Ponty, the concept of the natural world, of *physis*, is no longer something opposed to the human subject, in that

> we witness every minute the miracle of related experience, and yet nobody knows better than we do how this miracle is worked, for we are ourselves this network of relationships. The world and reason are not problematical. We may say, if we wish, that they are mysterious, but their mystery defines them: there can be no question of dispelling it by some 'solution', it is on the hither side of all solutions. True philosophy consists in relearning to look at the world.[12]

10 Gargani, A., *Crisi della ragione*, Einaudi, Turin 1976, p. 33.

11 Merleau-Ponty, M., *Phenomenology of Perception*, quot., pp. 44-45.

12 Therein, p. XXIII.

To relearn how to see the world means, first and foremost, to overcome classical antinomies, and subsequently to position oneself beyond any calculating or pre-constituted thought, in order to reach that interrelational and 'coalescent' worldly fabric – made up of self and world – where meaning is originally given, that being of implication and latency which is no longer man, nor world, but rather a veritative backdrop that 'covers', sustains, nourishes, yet does not define, assure, or guarantee.

Merleau-Ponty's philosophical radicalism, which liberates thought from classical prejudices, proves essential within the domain of interrogative reflection, insofar as it presents itself as a continuous search and an effort to push ever further back the original question of the event of meaning. Consequently, Merleau-Ponty neither reduces reality to an intellectualist eidetic grid nor to empirical facts, but addresses the problem of how meaning, a sense, is produced within reality when thought approaches it – that is, in other words, how a *physiognomy* of reality comes into being.

It is necessary to understand, beyond intellectualism and empiricism, how a certain content acquires form; how, more simply, a thought becomes an embodied thought, subjected to the risk of shadow, opacity, nonsense, and death. In light of this, the new definition of being – which can no longer conform to empirical or rational canons – is that of a mortal, transient being, one that disperses, that vanishes, that is never fully constituted, nor fundamentally constitutive.

§ 2. *On bodies: to pulsate, to beat, to endure*

From what has been said thus far, it is clear that, for Merleau-Ponty, the antinomy of realism and idealism is rendered meaningless unless one returns to the world of phenomena, unless the preconstituted categories of subject and object are subjected to a careful genetic investigation.

> Probably the chief gain from phenomenology is to have united extreme subjectivism and extreme objectivism in its notion of the world or of rationality. Rationality is precisely proportioned to the experiences in which it is disclosed. To say that there exists rationality is to say that perspectives blend, perceptions confirm each other, a meaning emerges. But it should not be set in a realm apart, transposed into absolute Spirit, or into a world in the realist sense. The phenomenological world is not pure being, but the sense which is revealed where the paths of my various experiences intersect, and also where my own and other people's intersect and engage each other like gears. It is thus inseparable from subjectivity and intersubjectivity, which find their unity when I either take up my past experiences in those of the present, or other people's in my own.[13]

Thus, the field of inquiry indicated by Merleau-Ponty, arising from the need to overcome the dichotomous impasse, is no longer that of the "constituting consciousness" or that of the "constituted consciousness", but that of our concrete *being-in-the-world.* In this sense, phenomenology goes beyond rationalistic idealism and empiricist materialism: it reveals to us that pre-thetic substratum in which both philosophical currents find their relative validity and are simultaneously transcended, awakening a primordial world – prior to the idealizations and objectifications of knowledge – a raw being that eludes any constitution and forms the original ground of our experience.

> If we confine ourselves to phenomena, the unity of the thing in perception is not arrived at by association, but is a condition of association, and as such precedes the delimitations which establish and verify it, and indeed precedes itself [...] The unity of the object is based on the foreshadowing of an imminent order which is about to spring upon us a reply to questions merely latent in the landscape. It solves a problem set only in the form of a vague feeling of uneasiness, it organizes elements which up to that moment did not belong to the same universe and which,

13 Therein, p. XXII.

> for that reason, as Kant said with profound insight, could not be associated.[14]

The abandonment of the classical polarities of subjectivism and objectivism, of consciousness and thing, thus leads us to the presence of that phenomenological-existential world which is not the actualization of a preexisting being, but the foundation of being itself.

Subjectivity is not, as idealism claims, a first principle. It refers to an existential and temporal structure of nature, prior to the distinction between the objective and the subjective. This is the world revealed to us, according to Merleau-Ponty, by the structure of perception; according to Husserl, by phenomenological description: an unfinished and ambiguous world that logical discourse, by contrast, conceives as complete, closed to the possible, self-satisfied, autonomous, and consequently rationally defined.

The perceptual experience of phenomena situates us – and places us – within that pre-objective and non-thetic ground, prior to predicative operations and constitutive of them, which is the foundational site of our coexistence and '*coalescence*' (as Merleau-Ponty says) with things. Human experience must therefore be grasped in its intimate becoming, there where the distinctions of intellect – subject and object, self and other – have not yet intervened; it must thus be understood as lived experience.

> Perception is not a science of the world, it is not even an act, a deliberate taking up of a position; it is the background from which all acts stand out, and is presupposed by them. The world is not an object such that I have in my possession the law of its making; it is the natural setting of, and field for, all my thoughts and all my explicit perceptions. Truth does not 'inhabit' only 'the inner man', or more accurately, there is no inner man, man is in the world, and only in the world does he know himself. When I

14 Therein, pp.19-20.

> return to myself from an excursion into the realm of dogmatic common sense or of science, I find, not a source of intrinsic truth, but a subject destined to the world.[15]

For this reason, in opposition to intellectualism, dogmatism, and theoreticism, Merleau-Ponty affirms the irreducibility and specificity of lived experience: "The world is not what I think, but what I live through. I am open to the world, I have no doubt that I am in communication with it, but I do not possess it; it is inexhaustible."[16] Perception becomes that intra-worldly fabric in which we and the world are, that phenomenal field in which logos is given, meaning is given, the world is given, and further, that *veritative backdrop* – that is, the ground that underpins all knowledge – from which our reflections subsequently detach. The criterion of truth, in fact, is not provided by the content of perception, but by perception itself, by the act of consciousness that perceives and is perceived.

From what has been said thus far, Merleau-Ponty's aim is not to arrive at a more radical foundation of objective knowledge, nor at a comprehensive clarification of its presuppositions through the retrieval of pre-predicative forms. Rather, it is to bring forth the meaning of the worldly horizon, not as the object of reflection on the modalities of its givenness, but as the meaning of man's situation as an existing being, of his lived and operative factuality. Merleau-Ponty's phenomenological philosophy therefore does not seek to explain the world or to discover its conditions of possibility, but to formulate an experience of the world, a contact with the world that precedes all thought about the world. For this reason, the return to phenomena proves to be the key to overcoming classical prejudices founded by rationalistic and empiricist theories.

Phenomenology, when approached phenomenologically, apprehends the world in its nascent state, as a world of

---

15 Therein, pp. XI-XII.

16 Therein, pp. XVIII – XIV.

significations, and recovers the existential, which had been sacrificed to science:

> I am not the outcome or the meeting-point of numerous causal agencies which determine my bodily or psychological make-up. I cannot conceive myself as nothing but a bit of the world, a mere object of biological, psychological or sociological investigation. I cannot shut myself up within the realm of science. All my knowledge of the world, even my scientific knowledge, is gained from my own particular point of view, or from some experience of the world without which the symbols of science would be meaningless.[17]

The entire universe of science, therefore, is built upon the lived world, and to rigorously evaluate science itself and understand its meaning, we must first awaken this experience of the world, of which it is a secondary expression. To awaken this experience means returning to the world prior to knowledge, with respect to which every scientific and abstract determination is secondary, signifying, and dependent. Thus, in *The Structure of Behavior*, Merleau-Ponty proposes the concept of the totality of the phenomenon in opposition to the atomistic theories of physiology and psychology, which he regards as outdated and insufficient. Indeed, according to the philosopher, the explanation of phenomena through their reduction to simpler elements, from whose sum they would supposedly result, is entirely illusory, because phenomena possess their own individuality, a meaning that surpasses the sum of their parts:

> In fact, the reflexes themselves are never blind processes: they adjust themselves to a 'direction' of the situation and express our orientation towards a 'behavioral setting' just as much as the action of the 'geographical setting' upon us. They trace out from a distance the structure of the object without waiting for its point-by-point stimulation. It is this global presence of the

17 Therein, p. IX.

> situation which gives a meaning to the partial stimuli and causes them to acquire importance, value or existence for the organism. The reflex does not arise from objective stimuli, but moves back towards them, and invests them with a meaning which they do not possess taken singly as psychological agents, but only when taken as a situation.[18]

The human organism is a whole that in some way precedes and generates its organs; it is a unitary center of organization, permeated by an internal teleology.

> Prior to stimuli and sensory contents, we must recognize a kind of inner diaphragm which determines, infinitely more than they do, what our reflexes and perceptions will be able to aim at in the world, the area of our possible operations, the scope of our life. [...] There is, then, a certain consistency in our 'world', relatively independent of stimuli, which refuses to allow us to treat being-in-the-world as a collection of reflexes – a certain energy in the pulsation of existence, relatively independent of our voluntary thoughts, which prevents us from treating it as an act of consciousness. It is because it is a pre-objective view that being-in-the-world can be distinguished from every third person process, from every modality of the *res extensa*, as from every *cogitatio*, from every first-person form of knowledge – and that it can effect the union of the 'psychic' and the 'physiological'.[19]

The age-old problem of the relationship between organ and function must be resolved in the sense that the function not only creates the organ, but in a certain way is the organ itself. The organ is inhabited by the function, it is as if its incarnation, and the function is always something new, which expresses itself in behavior; behavior must therefore not be explained through a plurality of separate causes but understood. To understand means, ultimately, to describe without resorting to explanatory elements that go beyond what results from the simple observation of the phenomenon. It is not, therefore, an

18 Therein, pp. 91-92.
19 Therein, p. 92.

explanation of the higher by means of the lower – the vital by the physio-chemical, or the psychological by the physiological – but neither of the lower by means of the higher.

Behavior, according to the French philosopher, is the living transcendence of the alternative between the in-itself and the for-itself; it is the unfolding of an internal "possibility" of an organism. The emergence of behavior is the breaking of the in-itself, that is, of objective space and time, in which physical events unfold, such that each moment occupies only a single point in time, and it is the establishment of a new dimension, which experiences time as the continuous reprise of the past and anticipation of the future.

In his 1942 work, Merleau-Ponty asserts that behavior, inasmuch as it has a structure, does not belong to either of these two orders, interior or exterior. Behavior, taken globally, excludes the thesis of a pilot-consciousness hidden behind a visible body, but it equally avoids turning the latter into a sort of automaton to which, in man, psychic and spiritual qualities would be added. The structure of behavior is *neither classifiable as thing nor as idea, and thus, in the experience of behavior, I effectively transcend the alternative of the for-itself and the in-itself, which is, moreover, the fruit of a flawed dialectic.*[20]

In this first work, Merleau-Ponty combats the fragmentary mechanism that regards the body moment by moment and point by point. Mechanistic and spiritualistic materialism are products of abstract thought, for they derive from the arbitrary hypostatization of particular elements of a given structure which, as such, is neither material nor spiritual, but original. To substantiate his argument, the philosopher thus presents the clinical experience of the "*membre-fantôme*" (at other times,

20 As for the controversy with Sartre concerning the conception of dialectics, see F. Fergnani's *Introduzione*, in *Il corpo vissuto*, Il Saggiatore, Milano 1979, pp. 35–51, as well as the following pages of this volume.

he examines, for the same purpose, the pathological cases of the aphasic).

The theory of the phantom limb fits perfectly within Merleau-Ponty's speculation, since the phenomenon of the phantom limb expresses our persistent openness to the actions of which the arm is capable, the preservation of that field of possible actions which was ours before the mutilation. "The refusal to accept the mutilation by the person who suffers it," writes Zani, "is the work of the '*je engagé*' in a specific physical and intersubjective environment, which continues to reach out toward its world despite the mutilations. The body is the means, the vehicle of this '*engagement*,' of this 'être-au-monde'; a body, of course, no longer considered as mere matter, but as a psycho-physical organism (from which also derives the rejection of the classical distinction between soul and body, especially in the Cartesian dualistic sense)."[21]

> The consciousness of the phantom limb remains, then, itself unclear. The man with one leg feels the missing limb in the same way as I feel keenly the existence of a friend who is, nevertheless, not before my eyes; he has not lost it because he continues to allow for it, just as Proust can recognize the death of his grandmother, yet without losing her, as long as he can keep her on the horizon of his life. The phantom arm is not a representation of the arm, but the ambivalent presence of an arm.[22]

The phantom limb is, then, like the repressed experience of a former present that refuses to become past. It is an ambiguous experience, for it is neither a memory in the common sense of the term, nor, on the other hand, a psychic illusion of which one is unaware. What Merleau-Ponty seeks to draw out from

21 Zani, L. *Fenomenologia dell'essere in Maurice Merleau-Ponty*, *Rivista di filosofia neoscolastica*, 1957, n. 49, p. 125 (my translation).

22 Merleau-Ponty, M., *Phenomenology of Perception*, quot., pp.93-94.

these analyses is the concept of the existential, which remains his constant focus and central concern:

> Man taken as a concrete being is not a psyche joined to an organism, but the movement to and fro of existence which at one time allows itself to take corporeal form and at others moves towards personal acts. Psychological motives and bodily occasions may overlap because there is not a single impulse in a living body which is entirely fortuitous in relation to psychic intentions, not a single mental act which has not found at least its germ or its general outline in physiological tendencies. It is never a question of the incomprehensible meeting of two causalities, nor of a collision between the order of causes and that of ends. But by an imperceptible twist an organic process issues into human behavior, an instinctive act changes direction and becomes a sentiment, or conversely a human act becomes torpid and is continued absent-mindedly in the form of a reflex.[23]

Thus, the subject's openness to the world through the body it inhabits, the inherence of consciousness in one's own body, the concordance between the perception my body has of itself and the perceptual experience of the external object, the connection between the existence of a commonly perceived world and the co-presence of subjects – all become the essential threads interwoven in that extended discourse on the body and intercorporeity, on the subject and intersubjectivity, which constitutes the core of Merleau-Ponty's philosophy. In doing so, it subverts classical objectivity and celebrates the acknowledgment of an opaque and anonymous background that envelops and reabsorbs every situation within itself.

> Not only is the permanence of my body not a particular case of the permanence of external objects in the world, but the second cannot be understood except through the first: not only is the perspective of my body not a particular case of that of objects, but furthermore the presentation of objects in perspective cannot be understood except through the resistance of my body to

23 Therein, pp. 101-102.

> all variation of perspective. If objects may never show me more than one of their facets, this is because I am myself in a certain place from which I see them and which I cannot see. If nevertheless I believe in the existence of their hidden sides and equally in a world which embraces them all and co-exists with them, I do so in so far as my body, always present for me, and yet involved with them in so many objective relationships, sustains their co-existence with it and communicates to them all the pulse of its duration. Thus the permanence of one's own body, if only classical psychology had analyzed it, might have led it to the body no longer conceived as an object of the world, but as our means of communication with it, to the world no longer conceived as a collection of determinate objects, but as the horizon latent in all our experience and itself ever-present and anterior to every determining thought.[24]

Through the notion of the body – not understood in a physiological or psychological sense, but as presence in the world – Merleau-Ponty reaches a crucial juncture: the point from which he can not only effectively demonstrate the insubstantiality of traditional categories, which are thereby dismantled, but also bring to light the themes closest to his concern.

24 Therein, p. 106.

PART TWO
RAYS OF THE WORLD

CHAPTER 1
THE BEING-IN-THE-WORLD

§ 1. *The anchoring to the world through the body*

The reflection on the reality of the body, with which the previous chapter concludes, is, without doubt, the guiding thread of Merleau-Ponty's inquiry. Indeed, it is precisely from this reflection that the philosopher derives and demonstrates the fundamental premises of his thought.

The body, according to Merleau-Ponty, expresses, first and foremost, the impossibility of grasping the things of the world – and ourselves – from the outside, from that position or point "archimédique" which constitutes the privileged locus of objectivation. It is thanks to the body – insofar as it is situated, that is, always already in situation – that we discover there exists no knowledge of reality as survol, panoramic and panoptic, but rather an oblique, perspectival, and lateral knowledge.

The rootedness of the body in the world therefore surpasses both subjectivist and objectivist stances, situating itself within that ambiguous terrain which belongs properly neither to "things purely as things," nor to the 'knowing subject':

> The experience of our own body, on the other hand, reveals to us an ambiguous mode of existing. If I try to think of it as a cluster of third person processes – 'sight', 'motility', 'sexuality' – I observe that these 'functions' cannot be interrelated, and related to the external world, by causal connections, they are all obscurely drawn together and mutually implied in a unique drama. Therefore, the body is not an object. For the same rea-

> son, my awareness of it is not a thought, that is to say, I cannot take it to pieces and reform it to make a clear idea. Its unity is always implicit and vague [...] I have no means of knowing the human body other than that of living it, which means taking up on my own account the drama which is being played out in it, and losing myself in it [...] Thus experience of one's own body runs counter to the reflective procedure which detaches subject and object from each other, and which gives us only the thought about the body, or the body as an idea, and not the experience of the body or the body in reality.[1]

It is then evident how, for Merleau-Ponty, the ambiguity of the body is not the irreducible duplicity between *body-object* – as conceived by science or even by common sense – and *body-subject*, as it is lived and designated in everyday speech through the first person, but rather the double and inseparable ambivalence intrinsic to the *body-subject* itself. The *body-subject* is the phenomenal body which, even when becoming the body *object* of consciousness, is never a thing, and even when it is the consciousness *of* the object-body, is never an idea.

As body *object* of consciousness, it is not a living-thing or a thing-alive, nor something that can be objectively defined through the attribute of life, but rather a *living-lived.* As consciousness *of* the object-body, it is not an abstract thought, a conceptual representation that can be linguistically formulated and communicated, but the chiaroscuro expression of our life as original affectivity – in short, a *lived-thought.*

The phenomenal body is therefore "ambiguous," because it is the inseparable solidarity of the two ambivalences – *living-lived* and *lived-thought* – between which the *lived* itself functions as a hinge. The play of this double ambivalence constitutes the very movement of understanding: the never remaining on this side of the originary dimension of sense,

---

1 Merleau-Ponty, M., *Phenomenology of Perception*, quot., pp. 230-231.

and the never detaching meaning, even in the utmost distance of its most adventurous elaborations, from the lived matrix of sense.

Therefore, the phenomenality of the body is not merely the thought *transcendental belonging* of the *reduced lived experience*, but also the pre-reflectively lived *natural corporeity*.

Merleau-Ponty's phenomenology thus restores consciousness to life: it abandons the theory of the *body-object* and recovers that of the *body-phenomenon*, so that the *living-body* may again be understood in its history, in its realization within the horizon of meaning to which it belongs.

On the one hand, then, consciousness no longer appears as constituting consciousness or as pure being-for-itself, but as *perceptual consciousness*, as a *subject of behavior*. On the other hand, the body loses those classical connotations attributed to it by physiology. That is, the body no longer appears as an edifice of molecules or an aggregate of cells – as an object in the world or a mere juxtaposition of organs in space – but as the *medium* of our communication with the world, as the latent horizon of all our experience.

After all, even in its descriptions of the body, classical psychology had already ascribed to it characteristics incompatible with the status of an object.

> In the first place it was stated that my body is distinguishable from the table or the lamp in that I can turn away from the latter whereas my body is constantly perceived. It is therefore an object which does not leave me. But in that case is it still an object? If the object is an invariable structure, it is not one in spite of the changes of perspective, but in that change or through it. It is not the case that ever-renewed perspectives simply provide it with opportunities of displaying its permanence, and with contingent ways of presenting itself to us. It is an object, which means that it is standing in front of us, only because it is observable: situated, that is to say, directly under our hand or gaze, indivisibly over-

> thrown and re-integrated with every movement they make. Otherwise, it would be true like an idea and not present like a thing[2].

In particular,

> it is true that an object is an object only in so far as it can be moved away from me, and ultimately disappear from my field of vision. Its presence is such that it entails a possible absence. Now the permanence of my own body is entirely different in kind: it is not at the extremity of some indefinite exploration; it defies exploration and is always presented to me from the same angle. Its permanence is not a permanence in the world, but a permanence on my part. To say that it is always near me, always there for me, is to say that it is never really in front of me, that I cannot array it before my eyes, that it remains marginal to all my perceptions, that it is with me. [...] I observe external objects with my body, I handle them, examine them, walk round them, but my body itself is a thing which I do not observe: in order to be able to do so, I should need the use of a second body which itself would be unobservable.[3]

My body, insofar as it sees or touches the world, seems to occupy a privileged position; yet my skin is but a fold of that very same skin which constitutes the entire world. *Corporeity* is thus the primordial expression of our *being-in-the-world*, insofar as it is precisely in the inherence of the body to the world (*worldly body*) and of the world to the body (*corporeal world*) that meaning is given.

As one can see, it is precisely the corporeal interrelation between *I* and *world* that allows Merleau-Ponty to overcome both the subjectivistic and objectivistic standpoints, as well as the idealistic and empiricist positions analyzed in the second chapter. Indeed, if we were to halt at objective thought, we would find ourselves faced with the antinomy of *being-in-itself* (which is that of objects unfolded in space) and *being-*

2 Therein, p.103.
3 Therein, pp.103-104.

for-itself (which belongs to consciousness). The *Other* would then stand before me as an *in-itself* and yet exist *for-itself*. The *Other*, that is, in order to be perceived, would require from me a contradictory operation, since I would have to distinguish it from myself—thus situating it within the world of objects—and at the same time think of it as consciousness.

In objective thought, therefore, there is no place for the *Other* or for a plurality of consciousnesses; in fact, everything is reabsorbed into a sort of all-encompassing consciousness. Conversely, through his deepening of the theme of corporeity, Merleau-Ponty arrives at a fundamental insight: the *object of knowing* does not belong to the sphere of mere contents, definable *in itself a parte objecti*; nor is the *subject of knowing* the pure gnoseological subject, definable *a parte subjecti*. The one exists only in relation to the other. But not as two poles or two moments that come to meet in the cognitive relation – for that would still remain within the dogmatic and formal framework of objective thought – but rather as two terms that, lacking any independent determination, define themselves solely through their reciprocal relation.

Knowledge must therefore be interpreted not as the *adequation of two polarities* (*adaequatio intellectus rei*), but as the *event of a relation*, the point of intersection of which is the body.

The moment of correlation can be actualized only by the body, which, by *living* its contents, makes them its own, realizing itself at once in a synthesis that is genuinely enacted and not surreptitiously presupposed.

Thus, body and things weave together a system of relations without either constituting a *primum* with respect to the other; the determination *in itself* of these two elements is, in principle, impossible, and knowing, as *concrete knowing*, occurs only as recognition of their relativity and complementarity. Indeed, the *primum*, if there is one, is the *relation itself*, from which alone subject and object come into being. Only within the realm of abstract speculation can one speak of an *I* and of *things*; concretely, there exists only an *I* that knows things by *living them* and *appropriating them through its own body*, for it is, in its essence, *body* and *makes body with them*.

In this sense, our body, according to Merleau-Ponty, forms part of the very same fabric of the world.

> A woman may, without any calculation, keep a safe distance between the feather in her hat and things which might break it off. She feels where the feather is just as we feel where our hand is.[…] The blind man's stick has ceased to be an object for him and is no longer perceived for itself; its point has become an area of sensitivity, extending the scope and active radius of touch, and providing a parallel to sight.[4]

Therefore, the thing and the Other come to be the correlate of my body and, more generally, of my existence – of which my body is the stabilized structure. The *thing* is constituted through the grasp that my body takes upon it, and not as a meaning for an intellect; it is a structure accessible to the inspection of the body and presents itself to perceptual experience already charged with anthropological predicates. Since the relations among things are always mediated by our body, the whole of nature becomes the staging of our life. That is why, in the final analysis – according to the philosopher – once the mythology of the *in-itself* has fallen away, "we cannot conceive anything which is not perceived or perceptible"[5].

Subject and object, consciousness and world, body and things thus undergo a radical reinterpretation: the synthesis of the object is realized through the synthesis of the *own body* (*corps propre*), and this synthesis does not take place between two external elements – one subject and the other object – but is accomplished at every instant within the relational movement of existence.

4 Therein, p. 165.
5 Cf. therein, p. 373.

Moreover, even knowledge itself is thereby completely redefined: we know not *in spite of* our body, but *by virtue of* our body; our finitude is precisely that by which we come to know.

Finally, since it is the body that allows me to reach the *pre-thetic* dimension, one could not truly grasp the radicalism of Merleau-Ponty without first disentangling the body from its identification with consciousness. It is the body that draws along the intentional threads binding it to the surrounding world; as already noted, it is the *latent horizon* of our experience, continuously present prior to any determinate thought, and it is this body that reveals to us both the perceiving subject and the perceived world alike.

> System of motor or perceptual powers, our body is not an object for an 'I think', it is a grouping of lived-through meanings which moves towards its equilibrium. Sometimes a new cluster of meanings is formed; our former movements are integrated into a fresh motor entity, the first visual data into a fresh sensory entity, our natural powers suddenly come together in a richer meaning, which hitherto has been merely foreshadowed in our perceptual or practical field, and which has made itself felt in our experience by no more than a certain lack, and which by its coming suddenly reshuffles the elements of our equilibrium and fulfils our blind expectation.[6]

In the deepest strata of the body, we encounter the perceptive and sub-perceptive life that binds the body to pre-individual sentience – to the physiological and somatic life of others – a life that sinks its roots into sexuality, nourishment, and sleep, into a hidden and pulsating common sensing, a consenting and therefore a *co-existing*. Everybody, every *I*, is capable of sensing the other body, the other *I*, insofar as it, too, lives the life of sense, of dream, of wakefulness, of nourishment, in a shared participation within an originary horizon of meaning.

---

6 Therein, p.177.

The body, therefore, is not a duplicate of consciousness, but consciousness itself; it is not one among the means that allow us to relate to the world, but the *only* way of access to it. The phenomenological subject would not be consciousness if it did not inhere in a body – if it were not *body* – situated within a historical situation which it lives in the first person, as an originary and inalienable point of departure.

All this allows us to deny – contrary to what asserted by many critics[7] – the belonging of Merleau-Ponty to that consciencialism characteristic of much of the French philosophical tradition. For Merleau-Ponty, in fact, the radicality of the reduction consists precisely in becoming aware of our relation with the world, and it is the body that introduces us into the originary, antepredicative nature: "we are ourselves this network of relationships (…) True philosophy consists in relearning to look at the world"[8].

The philosopher, therefore, does not seek to withdraw from the world into a pure consciousness; rather, he inscribes consciousness within a world that is always radically primordial, within an original presence that is neither thematic nor thetic. Correspondingly, consciousness is never *bare consciousness*, nor is it ever a *thing*: it is always participation in the world through a body, within an existential relation whose very dimension is, therefore, its ambiguity – that is, its constitutive *duplicity*.

From what has been said thus far, it follows that for Merleau-Ponty *inherence in the world* is a fundamental and ineliminable given – it is our very mode of existing – for I live *in* the world and the world lives *in me*: it is, in a sense, a kind of *copulation* between myself and things, and between myself *and as* things.

---

7 Cf., for example, G. Scotti, *Sulla percezione in Merleau-Ponty*, "aut-aut", 1957, n. 41, p. 520.

8 Merleau-Ponty, M., *Phenomenology of Perception*, quot., p. XXIII.

The world is not deduced from the *I*, just as the *I* is not deduced from the world; I feel the presence of the world only insofar as I manage to free myself from every systematic premeditation – only insofar as I can accomplish that *phenomenological reduction* which, for Merleau-Ponty, leads to the concreteness of perception—only insofar as I truly manage *to feel* and to allow the world *to feel and to feel itself within me.* If I wished to translate perceptual experience, I should say that *it perceives in me*, and not that *I perceive.*

The world perceives itself in me because its level differentiates itself, transcends itself, and opens itself within me. Thus, the surrounding world, in its historical unfolding, announces itself within our bodily and sensible life – at the intersection of those experiences that constitute the entire intersubjective fabric. Our

> life has, in the astronomical sense of the word, an atmosphere: it is constantly enshrouded by those mists we call the sensible world or history, the on of the corporeal life and the one of the human life, the present and the past, as a pell-mell ensemble of bodies an minds, promiscuity of visages, words, actions, with, between them all, that cohesion which cannot be denied them since they are all differences, extreme divergencies of one same something.[9]

The inextricable implication here described entails that integral being does not present itself as a pure object set before me, but "intersection of my views and at the intersection of my views with those of the others, at the intersection of my acts and at the intersection of my acts with those of the others, makes the sensible world and the historical world be always intermundane spaces".[10]

And the very same implication that binds me to the perspectives of my contemporaries also binds me to those who are no longer with me, within a relation of *simultaneity*

9 Id., *The Visible and the Invisible,* quot., p. 84.
10 Ibid.

– a term that designates a dimensional dilation of the field of presence, in which the past is precisely "simultaneous" with the present, understood in its more restricted sense.

In place of a merely horizontal or serial history, there thus takes shape the sense of a *spatio-temporal flesh*, wherein the simultaneous and the successive acquire significance together – where, in other words, *synchronicity* extends over *diachrony* without effacing it. In this *"vertical history"*, or history of *accumulation*,

> we are experiences, that is, thoughts that feel behind themselves the weight of the space, the time, the very Being they think, and which therefore do not hold under their gaze a serial space and time nor the pure idea of series, but have about themselves a time and a space that exist by piling up, by proliferation, by encroachment, by promiscuity – a perpetual pregnancy, perpetual parturition, generativity and generality, brute essence and brute existence, which are the nodes and antinodes of the same ontological vibration.[11]

Once again, Merleau-Ponty cautions us: there is no meaning already inscribed in the world independently of our embodied presence; there are no a priori universals dictating a path or an end to history, nor any pre-written movement, any law traversing the actions of individuals and peoples to assert itself as the red thread of history. Yet neither can worldly history be conceived as a cosmos that we – *Copernican* subjects – have constructed and continue to construct upon a fundamental void of meaning, upon a radical non-sense. Against this alternative – between objective meaning and radical non-meaning – Merleau-Ponty suggests that existence itself, as adherence and inherence to the world, is already the having-and-giving of meaning: *we confer upon history its significance, but not without its putting that significance forward itself*[12].

11 Therein, p. 115.

12 Id., *Phenomenoogy of Perception*, quot., p. 522.

And this meaning to which we are destined is, from the very beginning, communication, for solipsism is impossible, and I am always already grafted onto others.

Être-au-monde is thus synonymous with a sense that ignites, with a fabric that begins to weave itself—without this presupposing a pre-given pattern of essences, or my having to consider myself as the sole weaver of the entire tapestry.

Within this same theoretical framework, Merleau-Ponty therefore confronts the problem of alterity. It is always through his corporeal existence, through his being-situated, that the human being opens himself to communication with the other, for: "man finds himself bound to the other man from the very moment perception allows him to sense that he is *different* from them. From that very first moment, he must recognize that if he exists, it is because the *others*, acting as a mirror, reflect his own image back to him. Reflection will come later – together with reasoning and all the errors and unilateralities it entails – but if we wish to return to that primordial moment of the purest contact with things, we cannot go astray: the *others* are there, with us, and we are consciously *here* because the others are (…) The *Other*, then, becomes the very condition of existence for every *I*. My existence moves within the others even before I become aware of it"[13].

> Through the action of culture, I take up my dwelling in lives which are not mine. I confront them, I make one known to the other, I make them equally possible in an order of truth, I make myself responsible for all of them, and I create a universal life. Just as by the thick and living presence of my body, in one fell swoop I take up my dwelling in space. And like the functioning of the body, that of words or paintings remains obscure to me. The words, lines, and colors which express me come out of me as gestures. They are torn from me by what I want to say as my gestures are by what I want to do. In this sense, there is in all ex-

13 Ferraris, A. *L'apertura all'alterità nella filosofia di Merleau-Ponty*, "Atti dell'Accademia delle Scienze di Torino", 1962-63, pp. 261-263 (my translation).

> pression a spontaneity which will not tolerate any commands, not even those which I would like to give to myself. Words, even in the art of prose, carry the speaker and the hearer into a common universe by drawing both toward a new signification through their power to designate in excess of their accepted definition, through the muffled life they have led and continue to lead in us, and through what Ponge appropriately called their "semantic thickness" and Sartre their "signifying soil.". This spontaneity of language which unites us is not a command, and the history which it establishes is not an external idol: it is ourselves with our roots, our growth, and, as we say, the fruits of our toil.
>
> Perception, history, expression – it is only by bringing together these three problems that we can put Malraux' analyses in their proper perspective.[14] We are in the field of history as we are in the field of language or existence.[15]

My *being-situated*, which at first glance may appear as the destruction of every certain and definitive truth, reveals itself instead as the path toward a *truth-in-situation* – one that excludes the very possibility of perfect reciprocity and transparency. Situated finitude indeed entails also the reverse side of pure communication: the opacity that regenerates itself at every step, the shadow that accompanies each of our movements.

It thus carries within itself the possibility – and, to a certain extent, the inevitability – of a failure of communication, of an unbridgeable difference, of the impossibility of identifying oneself with the Other, of the emergence of *incommunicability* or *discord*: the Other withdraws into an incomprehensible transcendence. And yet, the Other, despite everything – in his strangeness, in his very opacity – continues to occupy a place within my horizon, and I cannot dispense with the problem of communication, since the intelligible structure of difference, as that of discord, has its reverse in a fabric that cannot be torn to the point that its terms lose all contact.

---

14 Merleau-Ponty, M., *Signs*, quot., p. 75.
15 Therein, p. 20.

> Conflict – writes Merleau-Ponty – is possible only on the basis of a primitive commonality. Two absolutely separate worlds are incommensurable; they exist in a transcendence as profound and impossible (or non-sensed) as the transcendence of a wholly other future world.[16]

The reflection on the body – as the domain that allows us to apprehend ourselves within the originary fabric of the world in which we are and of which we are – thus leads Merleau-Ponty to broaden his perspective in favor of corporeality. It is no coincidence that, in a self-presentation written in 1954, Merleau-Ponty wrote: “These observations [the ones made in *Phénoménologie de la perception*] lead us toward new investigations, which must definitively establish the philosophical meaning of the preceding ones, and which, in turn, prescribe to the new ones an itinerary and a method.”[17]. It is here, then, that one can grasp the progressive shift from the phenomenological–humanistic domain to the *ontological–carnal* one, marked by the passage from *corps propre* → *chair du monde* (*own body* → *flesh of the world*).

Yet the premises of this transition – testifying that it is, in fact, a continuous deepening of those very nuclei of thought that will evolve in the direction of ontology – are already contained within Merleau-Ponty’s first work.

My body, the natural world, space, the human world, and time must not be considered under the aspect of the physical sciences, as determined objects, *partes extra partes*, but must be apprehended within the system *moi–autrui–les choses* in its nascent state. This system reveals the ancient pact, concluded at birth, between the *I* and the world—*without me or world, but not without me and not without the world* – wherein a system of anonymous functions operates, functions already accomplished and already present.

---

16 See Id., *Sense and Non-sense,* quot., pp.63-70.

17 Id., *Autopresentazione*, “aut-aut”, n. 232-233, 1989, p. 8 (my translation).

The meaning of spatial orientation, the vision of space in depth, the grasp of movement and magnitude that inhabit things, make manifest those shifts of situational level through which the body takes differing distances from the world.

These experiences, apprehended in their nascent state, make the body appear as the *power through which intramundane relations are effected*, through a communication more ancient than thought itself.

Thus, we are placed immediately and always already within this worldly horizon, in which the ceaseless presence of our body constitutes both *adherence to* and *interrelation with* the world – a primordial horizon that is nothing other than the coupling of our body with things.

From all this one understands the importance of the theme of *perception* as the very domain from which our original relation with the world is given.

§ 2. *Anonymous perception in the world of the "one"*

The worldly domain of perception, with which the preceding paragraph concluded, thus reveals itself as the cradle of meanings, the sense of all senses, and the ground of all thought. Indeed, Merleau-Ponty himself, from the very beginning and in reiterating the central assumptions of his philosophy, emphasizes that the organization of the perceptual whole (and consequently its articulation into "things") is a primary fact – one that cannot be deduced from a multiplicity of elementary particles or from mere sensory aggregates. In both cases, as Bonomi observes, "What is lost from view is the concrete perceptual experience, the indissoluble connection between the body and the world"[18]. Perception is not mere

18 Bonomi, A. *Esistenza e struttura. Saggio su Merleau-Ponty,* quot., p.74 (my translation).

intellection; it is always mediated by the body, which is at once *contingency* and *transcendence*.

If we did not *have* a body – if we were not *body* among *bodies* – perception would never take place. This is why, according to Merleau-Ponty, *against the notion of a constituting consciousness that posits its objects as meanings, it is necessary to reclaim the facticity of the external world*:

> The mind that perceives is an embodied mind, and it is the rooting of the mind in its body and in its world that we have first sought to reestablish – both against the doctrines that treat perception as the mere result of the action of external things upon our body, and against those that insist upon the autonomy of conscious awareness. These philosophies share a common feature: they forget, in favor of pure exteriority or pure interiority, the bodily insertion of the mind – the ambiguous relation we maintain with our body and, correlatively, with perceived things. (…) Now, if perception is thus the common act of all our motor and affective functions as much as of our sensory ones, we must rediscover the figure of the perceived world through a work comparable to that of the archaeologist, for it lies buried beneath the sediment of subsequent knowledge. We would then see that sensible quality is not that opaque and indivisible datum presented as a spectacle to a distant consciousness, as classical conceptions maintained, and that colors, for example – each surrounded by an affective atmosphere that psychologists have been able to study and define – are in truth diverse modalities of our coexistence with the world. We would see that spatial figures or distances are not so much relations between various points of objective space as relations between them and a center of perspective that is our body – in short, that they are diverse ways in which external stimuli put to the test, solicit, and vary our grasp upon the world, our anchoring in the horizontal and the vertical of place, in a here and a now. (…) If we wish to define a subject capable of this perceptual experience, it is clear that it will not be a thought transparent to itself, absolutely present to itself, without body or history interposed. The subject of perception is not this absolute thinker; it functions through a pact concluded at our birth between our body and the world, between ourselves and our body. It is like a continued birth, the one to whom a physical and historical situation has been entrusted to manage – and this,

> at every moment anew. Each embodied subject is like an open register, the content of which remains unknown, or like a new language whose works we cannot foresee, but which, once it has appeared, cannot fail to say something – little or much – to have a history or a meaning.[19]

Perception thus reveals itself to be *communion*, *coexistence*, a lived and ambiguous *relation* between subjective and objective elements that enter the perceptual process as *intentional facts*. The synthesis is not accomplished by the *epistemological subject*, but by the *phenomenal body*, whose *intentionality* merges with the *psychophysiological unity* of the *bodily schema*.

Perceptual synthesis possesses no more the secret of the object than it does that of the body itself. The *perceived* presents itself as *presence–absence*, thereby rendering impossible the completion of synthesis in that *metaphysical point* which is the *thinking subject*. Indeed, what phenomenology identifies as the real unity of the thing is also that which withdraws it from our possession: the *irrecusable presence* and the *perpetual absence* of the object are the two inseparable aspects of its *transcendence*, which finds its counterpart in the subject as the *open and indefinite unity* of the bodily schema within the general movement of *being-in-the-world*.

Yet, since perceptual activity is intimately bound to the body, it will necessarily be *perspectival*, *lateral*, *partial* – it will appear as indefinitely *experienceable* and *explicable*, without, however, the synthesis of its various perspectives ever being completed: neither does the world ever give itself in its entirety, nor does the subject ever fully project itself into things.

Here, *ambiguity* with regard to the object means that it is grasped determinately only from a certain point of view; *ambiguity* with regard to the subject means that, in the act

19 Merleau-Ponty, M., *Un inedit,* in "Revue de Metaphisique et de Morale" n. 4, 1962, pp. 402-404 (my translation).

of consciousness, it is at once *coincidence with* and *distance from* the real – and that, by the very nature of its act, it must prolong itself into the future, exposing itself to *uncertainty, risk, failure*, and *error*. Consequently, a *total reflection* will never be attained precisely because the world gives itself only as a *presumptive totality*.

In this way, "The ipseity of the thing is never attained: every aspect of the thing that falls under our perception is but an invitation to perceive further; it is a momentary pause within the perceptual process. Were we to reach the thing itself, we would have already unfolded it entirely before us, stripped of all mystery. It would cease to exist as a thing at the very moment we believed we possessed it[20]. Therefore, that which grants reality to the thing is precisely that which withdraws it from our possession: *the aseity of the thing, its unchallengeable presence and the perpetual absence into which it withdraws, are two inseparable aspects of transcendence*[21].

Perception is the structure that most clearly reveals the expressive unity of the body, understood as the vehicle of my being-in-the-world. "is not a science, it does not posit the things with which science deals, it does not hold them at arm's length in order to observe them, but lives with them; it is the 'opinion' or the 'primary faith' which binds us to a world as to our native land, and the being of what is perceived is the antepredicative being towards which our whole existence is polarized"[22]. Once again, then, Merleau-Ponty affirms that the system of experience is not unfolded before me as if I were God, but lived by me from a certain point of view. I am not its disinterested spectator, but a participant, and my inherence in this point of view makes possible both the finitude of my

20 Escher Di Stefano, A., *Il radicalismo filosofico come esperienza del vissuto in Merleau-Ponty,* Loffredo, Naples 1981, p. 228 (my translation).

21 Merleau-Ponty, M., *Phenomenology of Perception*, quot., p. 271.

22 Therein, p. 375.

perception and, at the same time, its openness to the total world as the horizon of all perception: *The world is an open and indefinite unity in which I have my place*[23]. It is necessary, then, to seek a new status for perception and to demonstrate that it is neither a state nor a quality, nor the consciousness of a state or a quality. And at the same time, to show that the subject of perception is neither a thinker who records a quality, nor an inert domain that would be struck or modified by it, but rather a power that co-emerges with a certain context of existence. It becomes clear, then, that for Merleau-Ponty the subject of perception is, so to speak, "anonymous". "So, if I wanted to render precisely the perceptual experience, I ought to say that one perceives in me, and not that I perceive. Every sensation carries within it the germ of a dream or depersonalization"[24].

In short, I am no more conscious of being the true subject of my sensation than I am of being the true subject of my birth or of my death.

> Between my sensation and myself there stands always the thickness of some *primal acquisition* which prevents my experience from being clear of itself. I experience the sensation as a modality of a general existence, one already destined for a physical world and which runs through me without my being the cause of it.[25]

And furthermore: *Perception is always in the mode of the impersonal 'One'.*[26] This perceptual "One" thus foreshadows that inalienable presence of myself in the world and of the world in me, which constitutes being-in-the-world. From all that has been said so far, it is clear that the ontological discourse the philosopher sought to attain has here found its realization. Indeed, as Moreau suggests – "Man is not

23 Therein, p. 354.
24 Therein, p. 250.
25 Therein, p. 251.
26 Therein, p. 279.

in the world, the world is not in man, but according to a fundamental conception of phenomenology: man is being-in-the-world. The new notion of In-der-Welt-sein dissolves the age-old opposition between idealism and realism, and above all, it excludes the fundamental claim of idealist philosophy – namely, that of a spiritual activity radically distinct from the world and discovering within pure interiority a law that governs it, one that bears witness to the transcendence of the absolute and of value. […] The Heideggerian notion of being-in-the-world thus serves not only to settle the case between realism and idealism, but to abolish spiritual interiority and to evacuate ontological transcendence. […] The presence of the world is irrecusable. I am in the world, that is to say, not only is the world present to me, as idealism would have it, but its presence adheres to me: I cannot isolate myself from it; I am held within it and indissolubly present to it. I am not, properly speaking, contained or enclosed within it, as realism would have it, since it opens before me at the same time; yet I am present to it as it is present to me. This is further expressed by saying that I am situated in it – not situated as a point in space, but as a subject within the milieu in which its activity unfolds, a milieu in which it acts, yet from which it cannot be freed. I am situated in the world and in history, in the sense that I have a situation within society: this is what is meant by being-in-the-world"[27].

Consciousness adheres to the world; it neither subsumes it (as in intellectualism) nor is it merely juxtaposed to it (as in empiricism). *Being-in-the-world* means being in internal communication with it prior to any reflective act: it is the primitive communication between the embodied subject and the world, which forbids their separation into substances. Just as dialectical tension once designated the internal relatedness of the elements of an organic form, so too existence expresses

27 Moreau, J. *La presence du monde et l'ouverture a L'Etre,* "Les Etudes Philosophiques", 1959, pp. 425-427 (my translation).

the equally internal relatedness of the elements within the *perceived-perceiving* form.

In this latter case, indeed, we are dealing with an even more organic interiority, a still more intimate interrelation of its elements: there is achieved a complete participation of the existent in the existed, and vice versa, by virtue of which the embodied subject and the world are nothing other than the shifting poles of a single field of interpenetrating tensions. The conductor of these tensions is the body, the anchorage of interiority to a world.

It is clear that the subject and the world, thus interrelated, are not only not separate substances, but not even entities possessing any proper autonomy: they are poles of an existential tension and, as such, may even change their polarity: "The world is wholly inside and I am wholly outside myself. (…) As I contemplate the blue of the sky I am not set over against it as an a-cosmic subject (…) I abandon myself to it and plunge into this mystery, it 'thinks itself within me', I am the sky itself as it is drawn together and unified, and as it begins to exist for itself; my consciousness is saturated with this limitless blue"[28]. The body, a living knot of interiority and exteriority, sustains and realizes this kind of permutation between the respective situations of the subjective and the objective. It is, in fact, sensing at the same time that it is sensed; it is therefore a relation of itself to itself, and a bond between me and things. This latter, decisive transition – from the relation of self to self to the bond between myself and things – is made possible precisely through the *sensing–sensed* connection. Then, as Bayer rightly observes – "The *inside* and the *outside* are inseparable. The entire world is within me, and I am outside myself. The subject that I am, in all my concrete nature, is inseparable from *"this-body-that-is-mine"* and from

28 Merleau-Ponty, M., *Phenomenology of Perception,* quot., pp. 249 and 474

"this-world-that-is-here"[29]. Here re-surfaces the key point previously highlighted. The concept of inseparability from one's own body and from the body of the world (that is, from the worldly fabric) is, we must reiterate, not only the central nucleus of Merleau-Ponty's speculation, but also – indeed above all – the very threshold of the philosopher's thought that foreshadows the conception of the body as *diffused and expanded within the flesh of the world.*

At this juncture, the passage from phenomenology to ontology becomes inevitable. Yet before entering into the properly ontological domain, it is fitting, in order to complete the analysis thus far, to allude to a few additional problematic implications.

The reversible interweaving of *I–world* also manifests itself subtly within the gaze, since the decisive matter is not *to see oneself*, but rather *"to know oneself as the object of a gaze* "[30]. Thus, as Gambazzi wrote in a recent article, "The *I* does not derive from the subject's capacity to see, but from its condition of being *visible* to a gaze. *The Visible and the Invisible* and the *Working Notes* confront us with a primordial and originary visibility – with a *pre-existence of the gaze* that is always on this side, a visibility in which I am looked at not by the eye as an organ of vision, but by the gaze as a function of my field of original reversibility. Thus, within the gaze – even within the gaze of things – there operates a prior seeing, antecedent to any actual eye that may look upon me: a gaze that may proceed, for example, from a *shimmering sardine can floating on the sea* (an episode analyzed by J. Lacan)[31]: It was a small

---

29 Bayer, R. *Merleau-Ponty et l'existentialisme,* "Revue Philosophique de la France et de L'Etranger", 1962, n. 87 (my translation).

30 See Merleau-Ponty, Maurice. *The structure of Behavior,* quot.

31 Lacan, J. *Seminar book XI,* transl. by A. Sheridan, ed. by J.A. Miller. W.W. Norton & Company, New York, 1981, p. 95. See in particular the originally studies by Lanfredini R., *Emotion and affection between phenomenology and psychoanalysis:*

can – a sardine can. It floated there in the sun […] And Petit-Jean said to me: "*You see that can? Do you see it? Well, it doesn't see you*!" He found this incident highly amusing – I less so. I thought about it. Why did I find it less amusing than he? It's an interesting question. To begin with, if what Petit-Jean said to me, namely, that the can did not see me, had any meaning, it was because in a sense it was looking at me all the same. It was looking at me at the level of the point of light, the point at which everything that looks at me is situated — and I am not speaking metaphorically.[32]

These observations can be productively compared with several passages from one of Italo Calvino's last works, *Palomar*, in which the author – through an analytical spirit remarkably close to Merleau-Ponty's phenomenological approach – describes a singular, profoundly problematic experience: "The reflection on the sea forms when the sun begins to set. […] It is the hour when Mr. Palomar […] enters the water, moves away from the shore, and the reflection of the sun becomes a glittering sword on the water, stretching from the horizon all the way to him. Mr. Palomar swims within the sword – or rather, the sword always remains before him; with every stroke it withdraws, never allowing itself to be reached. […] And wherever Mr. Palomar moves, the tip of that sharp golden triangle is him; the sword follows, pointing at him like the hand of a clock whose pivot is the sun. *"It is a special homage that the sun pays to me personally,"* Mr.

behavior, body, memory, in "Chiasmi International", n. 18, 2016, pp. 33-48, on impersonal dimension into human consciousness, dimension analyzed from the dual viewpoint of phenomenology and psychanalisis. See also: Bernet R., *The Phenomenon of the Gaze in Merleau-Ponty and Lacan,* in "Chiasmi International", n. 1, 1999, pp. 105-119; Letellier I., *Le réel et la non-altérité de l'autre. Lacan avec Merleau-Ponty,* in "Chiasmi International", n. 17, 2015, pp. 133-146.

32 Gambazzi, P., *Fenomenologia e psicoanalisi nell'ultimo Merleau-Ponty,* "aut-aut" 1989, n. 232-233, pp. 119-120 (my translation).

Palomar is tempted to think – or rather, the egocentric and megalomaniac self that lives within him is tempted to think so. But the depressive or self-destructive self that cohabits with the other in the same vessel objects: *"Everyone with eyes sees the reflection that follows them; the illusion of the senses and the mind keeps us all prisoners"*.

Then a third co-inhabitant, a more even-tempered self, intervenes: *"It means that, in any case, I am part of the sentient and thinking beings, capable of establishing a relationship with the sun's rays, and of interpreting and evaluating perceptions and illusions."* [...] *All this happens not on the sea, not in the sun,* thinks the swimmer Palomar, *but inside my head, in the circuits between my eyes and my brain. I am swimming in my own mind; it is only there that this sword of light exists.* [...] But he also thinks: *I cannot reach it; it is always there in front of me. It cannot be both inside me and something in which I swim; if I see it, I remain outside it, and it remains outside me.* [...] The sword exists only because he is there; if he were to leave, if all swimmers and sailors were to return to shore, or merely turn their backs to the sun—where would the sword go? In the dissolving world, the thing he would most like to preserve is also the most fragile: that marine bridge between his eyes and the setting sun. Mr. Palomar no longer feels like swimming; he is cold. Yet he continues: now he is bound to remain in the water until the sun disappears. [...] *What a relief it would be if he could annul his partial and doubtful self in the certainty of a principle from which all things derive!* [...] But within him remains a point where everything exists in another way – like a knot, a lump, a clog – the sensation that you are here but might not be; in a world that might not be but is. [...] If no eye – except the glassy one of the dead – were ever again to open upon the surface of the watery globe, the sword would no longer shine. [...] Mr. Palomar swims underwater; he surfaces – there is the sword! One day, an eye emerged from the sea, and the sword, which had already been there waiting for it, was finally able to display the slender sharpness of its point and its glittering brightness. They were made for one another, sword and eye – and perhaps it was not the birth of the eye that brought the sword into being, but rather the opposite, because the sword could not do without an eye to behold it at its tip"[33].

In Merleau-Ponty's terms, the être-au-monde finally opens itself to the *en-être*: we have now moved from phenomenology to ontology.

33 Calvino, I., *Palomar,* Einaudi, Turin 1985, pp. 15-20 (my translation).

CHAPTER 2
DWELLING IN SENSE

§. 1 *Politics, Institutions, and History*

> To be a philosopher is just one of many ways to exist, and, as Marx said, no one can flatter himself that he has exhausted ''religious existence," "political existence," "juridical existence", "artistic existence," or broadly speaking, "true human existence" in a "purely philosophical existence." But if the philosopher knows this […] if he forsakes the illusion of contemplating the totality of completed history and feels caught up in it like all other men and confronted by a future to be made, then philosophy fulfills itself by doing away with itself as isolated philosophy. This concrete thinking, which Marx calls "critique" to distinguish it from speculative philosophy, is what others propound under the name "existential philosophy".[1]

No completed history, no "*philosophie de survol*", no detached philosophy, but rather a militant philosophy, one that works and acts in the world, that adheres to things and situations, that establishes relations, connections, openings, meaning. This is the French philosopher's project from his earliest works; this is, as he himself writes in the essay *Notes on Machiavelli*, "the center of philosophy"[2]. If militant philosophy is what Merleau-Ponty recognizes in Hegel's *Phenomenology of Spirit* – see the essay entitled *Hegel's*

1 Merleau-Ponty, M., *Sense and Non-sense,* quot., p.133.
2 Id., *Signs*, p.153.

Existentialism[3] – then, already at this early stage, we can clearly understand what is at stake in the task he assigns to thought: to uncover the radicality and the originary sources of institutions as a permanent counter-power. In direct confrontation with events – or rather, with the occurrences of history – Merleau-Ponty addresses the delicate question of philosophical and political engagement. If philosophy and politics separate or break our inherence in the world, our sense of dwelling within it, then the failure of both is close at hand. Instead, as Merleau-Ponty's postwar writings demonstrate, "by leading us to question tirelessly and continuously the nature and essence of what must be understood by politics, Merleau-Ponty's texts in the aftermath of the war truly offer the initial point of departure for a political philosophy, if only through the debate they provoke, and they urge us to return to the constant parallel that Merleau-Ponty draws between phenomenology and critical philosophy"[4]. The continuous questioning of facts, of actions, and of programs "institutes – writes Merleau-Ponty in response to Sartre's letter of July 1953 – correspondence that would irrevocably seal the rupture between the two – between a fact and the judgment made about it, a kind of constitutive derailment that disarms the trap of the event, allowing its meaning to appear more clearly."[5] This seems the most incisive aspect of Merleau-Ponty's political thought – a thought that speaks directly to questions that remain central in today's global political landscape, and

3 See Merleau-Ponty, M., *Sense and Non-sense* quot., pp. 63-70.

4 Dodeman, C., *La philosophie militante de Merleau-Ponty*, OUSIA, Bruxelles 2022, p. 161 (my translation).

5 See *Sartre-Merleau-Ponty: les lettres d'une rupture*, ed. by F. Ewald, in "Magazine Litteraire", 320, 1994, p. 76 (my English translation), it. transl. by D. Calabrò, E. Lisciani-Petrini, *Sartre/Merleau-Ponty. Il carteggio della rottura*, "Micromega", 1, 1997, pp. 193-221, then *Sartre, Merleau-Ponty: le lettere della rottura*, in "aut aut", 381, 2019, p. 35. The issue, entirely devoted to the correspondence, was edited by Enrica Lisciani-Petrini and Raoul Kirchmayr.

that therefore attests to the enduring relevance of the French philosopher. At this point, however, it is appropriate to proceed with order and begin from a set of necessary questions: How is history configured through the concept of institution? And what does Merleau-Ponty mean when he speaks of history and of its register – always open, yet fruitfully opaque?[6]. These questions lead us into the theoretical program developed by the French philosopher within the context of the postwar political debate. It is, first of all, a matter of understanding the practical implications of political-social *engagement* in light of the recovery and reworking of Hegelian dialectics, understood as *hyperdialectic*. From this point onward, the correlation becomes evident between dialectical thinking as

---

6 To these questions, Enrica Lisciani-Petrini provides the most incisive response over the course of her research devoted to Merleau-Ponty. See in particular: Lisciani-Petrini, E., *La passione del mondo*. Saggio su Merleau-Ponty, ESI, Naples 2002; Id, *Risonanze. Ascolto, Corpo, Mondo*, Mimesis, Milan 2006.
From here also derive her proposals for a "philosophy of the impersonal": Id, *Fuori della persona*. L'impersonale in Bergson, Merleau-Ponty, Deleuze, "Daimon", 44, 2012, pp. 73–88; Id, *Hacia el sujeto impersonal*, "Pléyade", vol. 9, 2012, pp. 137–159; Id, *Verso il soggetto impersonale*, "Filosofia politica", vol. 1, 2012, pp. 39–50; Id., *Per un "Dérèglement de tous les sens". Merleau-Ponty: le sinestesie e l'impersonale,* in "Chiasmi International", n. 10, 2008, pp. 109-127; Id., *"La voix de personne". Au-delà des frontières du sujet,* in "Chiasmi International", n. 11, 2009, pp. 348-358. Finally, it is to Enrica Lisciani-Petrini that we owe the proposal of a true "instituting thought," developed from an extensive reconstruction and analysis of the concept of "Institution" in Merleau-Ponty: see in particular: E. Lisciani-Petrini, *Potenza dell'istituzione*, "Discipline filosofiche", 2, 2019, pp. 71–99. On Institution theme, see also: Hirose K., *La genèse du social et la donation charnelle du non-présentable. La portée de la notion d'institution chez Merleau-Ponty,* in "Chiasmi International", n. 12, 2010, pp. 333-345; Larison M., *Stiftung et pensée du social: a propos de phénoménologie merleau-pontienne de l'istitution,* in "Chiasmi International", n. 18, 2016, pp. 363-376; Vallier R., *Institution: the significance of Merleau-Ponty's 1954 course at the collège de France,* in "Chiasmi International", n. 7, 2005, pp. 281-303.

a permanent interrogation, as a renunciation of the claim to definitiveness and to the unveiling of truth, and the inquiry into the meaning of history as something that cannot, in any way, dispense with the acknowledgment of an insurmountable ambiguity: the irreducible gap between the order of facts and the order of values, between events and the judgment that is made of them.

Keeping this theoretical framework in mind, it will not be difficult to understand the underlying motivations that led Merleau-Ponty to offer an interpretation of history that takes decisive distance from the Husserlian *conscientialist* solution – history as "constitution" – in order to configure itself instead as the *double bind* that is formed in contact with contingency and through which we are always exposed to the possibility of error – history as "institution."

> Institution [means] that this body, these gestures, these words, this life becomes the retort or the inverse of mine, and that happens be yond what is willed, experienced, known, since words have more senses than we give to them. To love, placed before analytic and spectator consciousness, is only to want to be loved because the other [for this form of consciousness] is only a not – me, constituting – constituted, and because to love can be only to become oneself again. The instituting subject in vests itself, i.e., animates itself with another meaning, transforms itself by means of its love, i.e., succeeds in making a meaning which is transcendent to him dwell in his I think and in his body, as a meaning dwells in the book and the cultural object. Therefore the visible institution is only the support of a spirit of the institution, certainly not immutable (love changes constantly, like a separation), but whose very change is a reaction of the event upon the instituted. An instituted is necessary so that there is an open register, history.[7]

7 Merleau-Ponty, M., *Institution and Passivity*. Northwestern University Press, 2010. p.10. On the theme of institution in Merleau-Ponty, in addition to the works by Lisciani-Petrini, E. already mentioned, see also: Bech, J.M., *'Institution' at the Crossroad: Ontological Extension or Corporeal Restatement,* in

The movement of dialectical thought that traverses history is nothing other, for the French philosopher, than the "labor" – in the Hegelian sense of the term – inhabiting the living ambiguity of events. Far, therefore, from an anthropocentric, mechanistic, or teleological conception of history, Merleau-Ponty approaches, in many respects, Weber's vision of history, to which he attributes the value of a non-ideological approach – particularly regarding the analysis of the capitalist process of rationalization. Such an approach makes it possible to loosen the opposition between subject and object, spirit and matter, and to move toward a conception of history as a plurality of intelligible centers, of "symbolic matrices." This is explained with great clarity by the French philosopher in a dense passage from his 1955 work entitled *Adventures of the Dialectic*: "They appear at the point where man and the givens of nature or of the past meet, arising as symbolic matrices which have no pre-existence and which can, for a longer or a shorter time, influence history itself and then disappear, not by external forces but through an internal disintegration or because one of their secondary elements becomes predominant and changes their nature"[8]. For Merleau-Ponty, it is not a matter of inscribing history within a teleology of reason, nor – still less – of confining it within a fixed and absolute order of values

"Chiasmi International", 11, 2009, pp. 201–247; Esposito, R., *Per un pensiero istituente*, in "Discipline filosofiche", 2, 2019, pp. 9–30; Id., *Pensiero istituente. Tre paradigmi di ontologia politica*, Einaudi, Turin 2020, especially pp. 199–207; Id., *Istituzione*, il Mulino, Bologna 2021, especially pp. 52–57; Gléonec, A., *L'institution et la passivité. Lectures de Merleau-Ponty*, Millon, Grenoble 2017; Larison, M., *Stiftung et pensée du social: à propos de la phénoménologie merleau-pontienne de l'institution*, in "Chiasmi International", 18, 2016, pp. 363–376; Revel, J., *Un registre ouverte*, in "Esprit", 319, 2023, pp. 77–89, in particular the section titled *L'institution*, pp. 86–89 ; Fortuzzi, M., *Istituire il futuro. Politica, istituzione e simbolo nel pensiero di Merleau-Ponty*, Mimesis, Milano, 2023.

8 Merleau-Ponty, M., *Adventures of the Dialectic*, quot., p. 16.

that would lead toward an uncontested eschatology of truth. Rather, history must be thought as a *"field of symbolism,"* as an *"ambiguous texture," "torn by all contingencies,"* open to every possibility. Events thus interlock with one another, forming a *"structured"* totality with multiple centers and multiple points of access.

The direction of history, therefore, is not subject to the idea of a homogeneous system that would chart its trajectories, nor to the *hybris* of a *Cogito* that would unify, within a closed universe of meaning, the multiplicity of social articulations. Instead, it is established in a certain oblique *"solidarity."* This *"solidarity"* – the intrinsic bond among the various orders of reality (economic, political, religious, social) – reaffirms the plurality of historical planes; that is, the fact that each plane of society is a symbolic system correlated with all the others, within a totality that is never definitively completed, but always exposed to the possibility of fragmentation. There is, therefore, no single vector of history, because the meaning of history lies neither in things nor solely in active subjects, but ignites with our very *incarnation*. In light of what has been said, particular significance is assumed by the Monday Course of 1953–54, entitled *Materials for a Theory of History*[9], in which Merleau-Ponty states that "the concept of history must be disentangled from a number of confusions."[10] First of all, the philosopher must recognize that – with regard to history – an artificial division has been created "between the wisdom of an understanding"[11] and the blind course of events, whereas "there is no question of choosing between external events and the internal spirit of man, between history and the timeless."[12], because

9 Merleau-Ponty, M., *Themes from the Lectures at the Collège de France, 1952-1960*, Northwestern University Press, Evanston, Ill. 1970.

10 Therein, p. 27

11 Ibid.

12 Ibid.

> the question is rather to know if this concatenation of the problems is an indication of the simultaneity of their resolution or whether this convergence and recuperation exists only in the process of interrogation. The proper starting point is not from the alternatives of understanding and history or spirit and matter, but from those of history as an unknown god – the good or evil genius – and history as the milieu of life. History is a milieu of life if it can be said that there is, between theory and practice, between culture and man's labor, between epochs and individual lives, between planned actions and the time in which they mature, an affinity that is neither fortuitous nor grounded in an omnipotent logic. [...] There is no history where the course of events is a series of episodes without unity, or where it is a struggle already decided in the heaven of ideas. History is there where there is a logic within contingence, a reason within unreason, where there is a historical perception which, like perception in general, leaves in the background what cannot enter the foreground but seizes the lines of force as they are generated and actively leads their traces to a conclusion. This analogy should not be interpreted as a shameful organicism or finalism, but as a reference to the fact that all symbolic systems – perception, language, history – only become what they were although in order to do so they need to be taken up into human initiative.[13]

From the final words of this passage, a theme particularly dear to Merleau-Ponty emerges namely, the convergence of three symbolic systems – perception, history, and language. The sensible world, the historical world, and the linguistic world all point toward the same phenomenology: a "philosophy of history" is given only as the prolongation of the initial opening to the world and to others; and here, opening is already communication, perceptual adhesion. The inquiry into history is therefore inseparable from the inquiry into perception and expression, forming a valid and more elementary point of departure. Here lies, for Merleau-Ponty, the meaning of inhabiting the world and of assuming it from a political standpoint.

---

13 Therein, pp. 28-29.

The problem left unresolved by various philosophies of history – whether they rest upon external material or spiritual forces – is brought back to this side of both deterministic and consciousness-based visions of history: perception, history, and expression are not to be understood as three separate domains, but rather as "three symbolic systems." This implies that each system must be considered as a symbol of the other, and conversely.

The proposal advanced by Merleau-Ponty concerning the inherence and the tight co-implication of these three distinct fields of reflection is not – as it might at first appear – an elegant way out of the problem of history, but rather the sustained attempt—pursued throughout the entire arc of his philosophical work – to avoid reducing the plurality and complexity of historical events to a total solution, to a definitive synthesis, to a philosophy of survol that would guarantee, at any cost and by any means, the final realization of what had been *geometrically predetermined*[14]. It is not possible to transfigure the meaning of life and of human actions in the perspective of a homogeneous, contradiction-free society, for such a move would entail the total emptying out of the present, the annihilation of the past, and an illusory leap into the future. History continuously reformulates its own problems, without necessarily resolving them, without claiming to exhaust the meaning of being, without reference to a final endpoint. In his writings on painting and on literary language, Merleau-Ponty speaks of a "mute" or opaque historicity, one that solves problems only "obliquely", and for which "we see only what we look at"[15]. But it is precisely

14 A refusal to treat the problem of history as a *"geometrical problem"* is precisely what Claude Lefort has emphasized in *Sur une colonne absente. Écrits autour de Merleau-Ponty*, Gallimard, Paris 1978, particularly in the pages 45–115.

15 See Merleau-Ponty, M., *Eye and Mind*, in *The Merleau-Ponty Aesthetics Reader: Philosophy and Painting*. Northwestern University Press, Evanston, Ill. 1994, p. 124.; Id., *Signes*, quot., pp.

the impossibility of establishing a final balance that gives the story its true meaning: "The meaning of a system in its beginnings is like the pictorial meaning of a painting, which not so much directs the painter's movements but is the result of them and progresses with them"[16]. "For if we cannot establish a hierarchy of civilizations or speak of progress – neither in painting nor even elsewhere – it is not because some fate impedes us; it is, rather, because the very first painting in some sense went to the farthest reach of the future"[17].

We can therefore see how Merleau-Ponty rejects, in many places, the eschatological perspective of history, since such a perspective not only loses contact with the concreteness of life and of daily struggle, but also rushes blindly toward its presumed final goal, which becomes the object of a kind of blind faith – the faith in a hidden dialectical law that will, inevitably, produce its effects in the end, however disastrous they may be. From here arises the necessity of rethinking that dialectic which presents itself either as an internal (though invisible) objective structure of historical development, or which, on the contrary, grounds itself in pure subjective decisionism. Dialectic – the philosopher insists – is not a thought that constitutes the whole, but "belongs to it". It does not appear either in things or in subjects as such, but only "at the junction of a subject, of being, and of other subjects". Individual subjects are therefore released from the danger of solipsism, because the mode of Being proper to dialectic "is not only a spectacle that each subject presents to itself for its own benefit, but which is rather their common residence, the place of their exchange and of their reciprocal interpretation"[18]. This does not mean, however, that dialectic

---

49-104; Id., *L'institution dans l'histoire personnelle et publique*, in Id., *Résumés de Cours au Collège de France 1952-1960*, quot., pp. 59-65.

16 Id., *Adventures of the Dialectic,* quot., p.17

17 Id., *Eye and Mind*, quot., p. 149.

18 See Id., *Adventures of the Dialectic,* quot., pp. 203-204.

should be understood in an anthropological sense, for it is the *site of juncture* between subjects and Being. Dialectic neither grounds History nor Being; it is only *inscribed* within them. When, instead, it is detached from them and absolutized as the protagonist of History, it is immediately transformed into a power, into a kind of malign genius that acts in concealment, producing not only the risk of non-sense, "but much worse: the assurance that the things have *another sense* than that which we are in a position to recognize in them"[19].

> In thought and in history as in life the only surpassing's we know are concrete, partial, encumbered with survivals, saddled with deficits; there is no surpassing in all regards that would retain everything the preceding phases had acquired, mechanically add something more, and permit the ranking of the dialectical phases in a hierarchical order [...] But, on a defined part of the route, there can be progresses; especially there are solutions excluded in the long run. In other words, what we exclude from the dialectic is the idea of the pure negative, what we seek is a dialectical definition of being that [...] must rediscover the being that lies before the cleavage operated by reflection, about it, on its horizon, not outside of us and not in us, but there where the two movements cross.[20]

The central idea of Merleau-Ponty's reflection on history does not, therefore, depart from the notion of an *"open dialectical processuality"*, which characterizes the *"originary polymorphism of raw Being."* If we keep this connection in view, it will not be difficult to understand the rejection of a philosophy of history (and particularly of Sartre's) in favor of a *"philosophy of structure."* All of this undoubtedly derives from a marked interest in Saussurean linguistics, which, beginning in the 1950s, permeates Merleau-Ponty's analyses of perception, of the new ontological perspective, and of

19 Id., *The visible and the Invisible,* quot., p. 94.
20 Therein, p. 95.

history itself. This is what the philosopher writes in the *In praise of philosophy*:

> The theory of signs, as developed in linguistics, perhaps implies a conception of historical meaning which gets beyond the opposition of *things* versus *consciousness*. Living language is precisely that togetherness of thinking and thing which causes the difficulty. [...] The presence of the individual in the institution, and of the institution in the individual is evident in the case of linguistic change. [...] The contingent fact, taken over by the will to expression, becomes a new means of expression which takes its place, and bas a lasting sense in the history of this language. In such cases, there is a rationality in the contingent, a lived logic, a self-constitution of which we have definite need in trying to understand the union of contingency and meaning in history, and Saussure, the modern linguist, could have sketched a new philosophy of history.[21]

Phenomenology and structuralism here find themselves at a fruitful point of intersection: Merleau-Ponty identifies precisely in the idea of an *"anonymous and unconscious structure"* as a bond, a *"carnal junction"* with others and with the world, that which allows him to move beyond the *"various humanistic theories"* of a subject as the constructor of history. Language and history, insofar as they are symbolic systems, represent the succession of equilibria that disintegrate and reorganize themselves, of relations of communication that are confirmed only laterally, through trial and error. In this sense, the dimension of philosophy is that of coexistence; understood, however, "not as a fait accompli and an object

---

21 Id., *In Praise of Philosophy and Other Essays.* quot., pp. 54–55. Still on the question of *institution*, to which Merleau-Ponty refers in the passage cited above, see also *Résumés de cours au Collège de France*, pp. 59–65, where institution is understood as that which makes possible the enduring dimensions of our personal and public experience, in a *"rhythm of conservation, resumption, and surpassing"* (ibid., p. 61). See also A. Lingis, *Phenomenological Explanations*, M. Nijhoff, The Hague 1986, pp. 77, ff.

of contemplation, but as the milieu and perpetual event of the universal praxis"[22]. It is evident, from what has been said so far, that the parallel between the gap of the linguistic sign – which arises in the intervals between the words in which it is embedded – and the gap of historical meaning – represented by the distance between the fact (the thing) and the judgment (the consciousness) of the fact – refers us immediately to the essential ambiguity of Being; there, where "there is room, without contradiction and without magic, for relationships with double meanings for reversals, for opposite and inseparable truths, for sublations, for a perpetual genesis, for a plurality of levels or orders."[23] The reference to the plurality of orders indicates not only the irreducibly open and multidimensional sense of being, but also that of responses, which are always characterized by *partial* successes and *local* advances. It is not possible to speak of a definitive and global answer to the problem of history if it is conceived as a single, compact problem: disappointment awaits "whoever believed in salvation, and in a single means of salvation in all realms"[24].

§ 2. *Merleau-Ponty's Political Humanism*

Behind this conception of history one can discern the Merleau-Ponty of the 1940s – the Merleau-Ponty of *The Structure of Behavior* and *Phenomenology of Perception* – where philosophical reflection is organized around the theme of relational being, generating several essential concepts: the concreteness of the existential situation; consciousness rooted in the world; existence understood as the constitutive and structural co-presence of consciousness *with* the world and of the world *within* consciousness. In truth, this orientation led

22 Id., *Signs*, quot., p.112.
23 Id., *Adventures of the Dialectic*, quot., pp. 203-204.
24 Id., *Signs*, quot., p. 35.

the French philosopher to engage directly and personally in political life. For the affirmation of the primordial intertwining of human being and world – and therefore of intersubjectivity – could not but require an attentive reflection on society and its values.

At the same time, an interest in the thought of the young Marx was taking shape in France, appearing to respond to the expectations of the most influential philosophical currents of the time: existentialism, neo-Hegelianism, and various strands of Catholic thought. Thinkers of different intellectual backgrounds – Mounier and Sartre, Bigo and Hyppolite, Merleau-Ponty himself and Lefebvre – thus found themselves reading the early Marx, motivated on the one hand by the need to problematize traditional Marxism in the name of recovering the themes of the human – subjectivity and freedom – and, on the other hand, by the rediscovery in the early writings of what seemed an *authentic* Marxism, capable of being opposed to the already distorted and ossified versions of official doctrine.

Moreover, it became apparent that Marx's early texts could serve as a precise critical instrument against Stalinist orthodoxy; yet for that very reason it was necessary to uncover, in all its depth and implications, the philosophy implicitly at work within them. It is at this juncture that Merleau-Ponty's elaboration of historical becoming as a dimension of possibility and contingency begins to take shape. Indeed, in the characteristic tension between an authentic interpretation of Marxism and the deformed one of Stalinist orthodoxy, Merleau-Ponty not only frees historical materialism from the constraints of a philosophy of history governed by deterministic structures evolving into an impenetrable historical necessity but also dismantles the very presupposition that derived historical materialism from dialectical materialism.

Through the early writings of Marx, the French philosopher rethinks Marxism as a revolutionary project that posits *de-alienation as genuinely possible*, thereby emphasizing the dimensions of historical responsibility and of subjective awareness: “As soon as man is introduced as the subject of history – and Marxism so portrays him-one is no longer bringing in merely collective man or class but is also including individual man who retains his power to serve or to betray his class and who in this sense joins it of his own accord”[25].

All of this has far deeper origins, already traceable to the period in which Merleau-Ponty, in the *Phenomenology of Perception*, spoke of concrete subjectivity (corporeal and perceiving) as openness, as the privileged point of reference for every discourse and every action upon the world. And it is precisely this Merleau-Pontyan conception of the human being – grounded in the irreducible character of existence and in the necessity of a direct political engagement – that establishes the conditions for the encounter with Marxism[26]. But for

---

25 Id., *Sense and Non-sense*, quot., p.80.

26 On Merleau-Ponty’s long dialogue with Marxism, see: Fergnani, F., *Marxismo e filosofia dell’ambiguità*, “Il pensiero critico”, 4, 1960, pp. 16–65; Paci, E., *Merleau-Ponty, Lukács e il problema della dialettica*, “aut aut”, 66, 1961, pp. 498–516; Bonomi, A., *Esistenza e struttura. Saggio su Merleau-Ponty*, Il Saggiatore, Milan 1967, pp. 152–181; Rabil, A., *Merleau-Ponty, Existentialist of the Social World*, Columbia University Press, New York 1967; Moravia, S., *La crisi della generazione sartriana*, “Rivista di Filosofia”, 58, 1967, pp. 426–470; Pompeo Faracovi, O., *Eresia e riformismo in Merleau-Ponty*, “Il Ponte”, 9, 1972, pp. 1235–1259, and Id, *Filosofia e politica*, in VV.AA., *Esistenza, filosofia, politica*, edited by G. Invitto, Guida, Naples 1982, pp. 173–197; Borg, J.L., *Le marxisme dans la philosophie socio-politique de Merleau-Ponty*, “Revue Philosophique de Louvain”, 73, 1975, pp. 481–509; Brena, G.L., *Alla ricerca del marxismo: Maurice Merleau-Ponty*, Dedalo, Bari 1977; Delogu, A., *Né rivolta né rassegnazione. Saggio su Merleau-Ponty*, ETS, Pisa 1980; Eslin, J.C., *Critique de l’humanisme vertueux*, “Esprit”, 6, 1982, pp. 7–20; Marinotti, A., *Percezione e conoscenza. Linee della fenomenologia genetica by M. Merleau-Ponty*, Alfani, Florence 1983, especially chap. V, pp. 112–148; Invitto, G., *Merleau-Ponty: l’eresia programmatica*, Lacaita, Manduria 1971, and Id, *Merleau-Ponty e la filosofia come vigilanza*, Milella, Lecce 1981; Lefort, C., *Sur une colonne absente. Écrits autour de Merleau-Ponty*,

this to truly take place, it is necessary that Marxism renew itself; above all, it is necessary to recover within it precisely the dimension of concrete subjectivity, which had long been concealed by the theoretical orientations expressed by official Marxism. Merleau-Ponty thus elaborates a "Marxism of waiting", which represents a highly significant attempt to mediate between existentialism and Marxism – not with the intention of surpassing the latter, but rather of reclaiming it.

The French philosopher therefore tends to problematize the forms of contemporary Marxism, opposing to its "materialistic versions" the philosophy of the "true" Marx – that is, the Marx in whom one encounters a meaningful philosophy of praxis and a distinctive inclination toward all those ethical, subjectivist, and consciousness-centered motifs that orthodox Marxism had entirely marginalized.

Gallimard, Paris 1978, and Id, *D'un doute à l'autre*, "Esprit", 6, 1982, pp. 23–30, and Id., *Merleau-Ponty en accusation. Mise au point*, "Esprit", 10, 1985, pp. 83–86; Peillon, V., *La tradition de l'Esprit. Itinéraire de Maurice Merleau-Ponty*, Grasset, Paris 1994, in particular see chapters II and III of the second part and chapter I of the third part; Lisciani-Petrini, E., *Presentazione de Il carteggio della rottura*, "MicroMega", 1, 1997, pp. 193–198; Id., *La passione impolitica della politica. Merleau-Ponty tra filosofia e non-filosofia*, in VV.AA., *Nichilismo e politica*, Laterza, Bari 2000, pp. 55–73; Id., *Merleau-Ponty/Sartre: una insanabile divergenza filosofico-politica*, "aut aut", 381, 2019, pp. 61–90. To these last studies may be added those collected in the volume edited by Alloa E. and Revel J. devoted to the "Merleau-Ponty politico" in the journal "Chiasmi International", 19, 2018, especially: Dodeman, C., *Une "philosophie militante" est-elle encore possible?*, pp. 39–56; Belot, D., *"Cette idée d'une histoire comme champ... des Les Aventures de la dialectique à la Préface de Signes"*, pp. 57–68; Chouraqui, F., *The Paradox of Power*, pp. 69–85; Monferrand, F., *Politiser l'expérience. Merleau-Ponty, Socialisme ou Barbarie et l'expérience prolétarienne*, pp. 87–100; Alloa, E., *Merleau-Ponty à Madagascar. L'épreuve de l'étranger et la décolonisation de la pensée*, pp. 115–127.

And it is precisely here that – as Lisciani-Petrini masterfully explains – his divergence from Sartrean thought comes sharply to the fore.

> The *"thoughtful presence"* of which Merleau-Ponty speaks as the true *philosophical engagement with respect to politics* names nothing other than the inevitable *"strabismus"* to which the philosopher is necessarily constrained, precisely insofar as he wishes to remain a philosopher: that of looking *at once* from within and from without the world. From within – so as to be present to its political life, insofar as this world is his own historical dwelling, literally his *ethos*. From without – to discern more clearly the sense that, each time, takes shape within it, to grasp how and from where it arises. This is the practice of *epoché* – which, contrary to what Sartre believes, the philosopher not only can but *must* exercise in and toward the world in which he lives, so as to help sustain that self-reflection which every historical reality, like every position-taking, should never allow itself to abandon.
>
> For this reason, *"doing philosophy"* in this distanced manner and *"writing on politics"*—Merleau-Ponty can respond to Sartre – *"are one and the same thing."* […] Epoché, therefore, far from disengaging the philosopher from the world, is what places him at a distance so as to let its transcendences emerge; it loosens the intentional threads that bind us to the world so that these threads may be made visible. A "method closer to politics" than that of "continuous engagement" in "each event taken in itself" – Merleau-Ponty emphasizes to Sartre – precisely because politics is not the uninterrupted choice or intervention upon the immediate, but a vigilant reflection on the causes and consequences of each singular gesture or occurrence; a continual back-and-forth between present, past, and future, between the current singularity and the global complexity. […] Philosophical reflection and political thought measure themselves not by the contingencies of urgency, but by the underlying *dynamis* and the broader coordinates of each historical-worldly project.[27]

27 See Lisciani-Petrini, E., *"Le lettere della rottura" fra Sartre e Merleau-Ponty*, in *aut aut*, 381, 2019, pp. 9–20 (my translation), and again in the same author's *Merleau-Ponty/Sartre: una insanabile divergenza filosofico-politica*, cited above. There we find one of the fundamental passages that clarifies the opposed positions of

Merleau-Ponty thus returns to a well-known statement by Marx, who, in the *Ist Thesis on Feuerbach*, wrote: "The main thing wrong with al past materialism… is that it considers the thing, reality, the tangible world only as object or intuition, not as concrete human activity; as practice, not subjectively"[28]. The appeal to the *I*st *Thesis on Feuerbach*, understood as the affirmation of an ineliminable *subjective* dimension in Marx's thought, constitutes not only the basis for an explicit polemic against a form of Marxism that mechanically derives consciousness from things, but also the interpretative support for recovering the themes of individuality and subjectivity (a subjectivity that is an *"opening onto the world"*), and of consciousness as the grounding of reality and of history. From this there follows an attempt to reconstruct historical reality beginning from the dimension of subjectivity – an attempt that Merleau-Ponty strives to transpose into the very interior of Marxism itself. "Marxist praxis […] does not have the same resource. It cannot assign history a particular end in advance; it cannot even affirm the dogma of "total ' man" before he actually comes into being. If al our contradictions are someday to be resolved, then that day will be the first we know of it"[29]. And further on, Merleau-Ponty emphasizes again: "Marxism does not suppress history's subjective factors in favor of objective ones; it binds the two together."[30] From this particular vantage point – the need to preserve the role of subjectivity and consciousness in history, to rescue

---

the two philosophers with respect to the concepts of history and politics. Merleau-Ponty's diagnosis of *ultra-Bolshevism* in Sartre – formulated in response to Sartre's reproach that he fails to adhere to what Sartre considers the *"objectivity"* of history – reveals a position that, according to Merleau-Ponty, places philosophy itself and its very possibility in danger. The passage cited appears on pp. 89–90.

28 Merleau-Ponty, M., *Sense and non-sense*, quot., p. 80 (passage taken from "*Premiere These sur Feuerbach*", quot.

29 Therein, quot., p. 82.

30 Therein, quot., p.149.

them from the exclusive domain of objective determinations – the French philosopher finds in the pages of *History and Class Consciousness*[31] one of his most significant points of reference. Lukács's work in fact offers him the possibility of deepening, within the horizon of Marxism, the characteristic existential themes of contingency and, above all, of the ambiguity of the human condition and of history itself. From this perspective, the return to the young Marx does not mean merely attempting a refoundation of Marxism; it also implies the destabilization of every "scientific" discourse on history, of every vision that would conceive it in purely objective terms. Unlike theological philosophies, or even the systematic Hegel, it is indeed proper to Marxism to affirm that "humanity's return to order, the final synthesis, is not necessitated but depends upon a revolutionary act whose certainty is not guaranteed by any divine decree or by any metaphysical structure of the world"[32]: likewise, it should also be acknowledged that "history is both logical and contingent, that nothing is absolutely fortuitous but also that nothing is absolutely necessary".[33] If, therefore, one decisively gives up "the theological idea of a rational basis of the world, the logic of history becomes nothing more than one possibility among others"[34].

Merleau-Ponty thus anticipates a Marxism that, on the one hand, relinquishes the idea of an objective history, of a course of events that could be grasped in a scientific manner, and on the other hand, embraces the contingency and ambiguity that are intrinsic to human and subjective experience. Revolution and socialism then become only one among the infinite possibilities that stand open before human existence; the choice for revolution and for socialism – once any possibility

31 Lukacs, G., *History and Class Consciousness*, Merlin Press, 1967.
32 Merleau-Ponty, M., *Sense and non-sense*, quot., p. 120.
33 Ibid.
34 Therein, p.121.

of grounding them upon objective criteria has been lost – becomes an ethical option, a subjective moral need.

> The Marxist discovery of social existence as the most "interior" dimension of our life, of class dynamics as an integral process where economic and cultural determinations endlessly intersect and inter-signify, not only admits but demands a new conception of consciousness on the theoretical plane, which would establish a basis for both its autonomy and its dependence by describing it as a nothingness which comes into the world and which could not keep its liberty without engaging itself at every moment.[35]

The true sense of Marx's materialism thus consists in the effort to grasp the bond that links human being to the world, and human being to other human beings: matter and consciousness are situated within a single system of *coexistence*. Within this perspective, the concept of *praxis* becomes synonymous with social existence, with relation and engagement between the human being and the world, between the *I* and the *other*.

In this attempt to condition Marxism from within, Merleau-Ponty's own collaboration with the Communist Party and its political line took concrete shape. One had to carry out – according to the French philosopher – "the policy of the Communist Party" and "reconstruct with the proletariat." In November 1945, there was simply "nothing else to do"[36]. But Merleau-Ponty's theoretical articulation of Marxism did not at all coincide with the political positions expressed by the Parti Communist Français. The divergence – experienced by the philosopher in a genuinely dramatic way – finds its emblematic expression in *Humanism and Terror*[37], which is without doubt the culminating point of Merleau-Ponty's reflection in that period. This work constitutes, in fact, the testimony of an adhesion to Marxism conceived in the

35 Therein, p.82.
36 Therein, p.171.
37 Id., *Humanism and Terror.* Boston, Beacon Press, 1969.

only terms that could render it possible (that is, through the recovery of *praxis* as intersubjectivity), and at the same time one of the most significant documents of the French cultural and political climate of the time. *Humanism and Terror* contains within itself the seeds of the difficulties inherent in that task of "carrying out the actual politics of the Communist Party" that Merleau-Ponty had indicated already in 1945: critical allusions to the internal and international degeneration of communism, to the loss of its revolutionary spirit, to the *bureaucratized* revolution.

§ 3. *Reacting to violence and terror*

At the center of Merleau-Ponty's reflection now stands the problem of violence. This problem had already been widely debated, owing to the diffusion of Koestler's novels[38] and to the persistent anti-communist propaganda that insisted on the inhumanity of Soviet communism. With *Humanism and Terror*, the French philosopher attempts a solution that still, at the theoretical level, moves in a Marxist direction, by positing recourse to violence as an instrument for the necessary transformation of the existing order. Merleau-Ponty's position is once again an attempt at mediation: the ambiguity proper to history renders every choice perilous; one cannot simply be anti-communist, one cannot simply be communist. The only possible stance is one of critical waiting – a posture of proximity to and provocation of Marxism, in a moment in which one witnesses the "stagnation" of history and the "masking" of revolution in the world. Yet the matter does not end here. In the work of 1947, just as much – and perhaps more insistently – than the explicitly political theme of a "Marxism of waiting," one finds Merleau-Ponty's sustained

38 Koestler, A., *Darkness at Noon*, MacMillan, London 1940; and Id., *The Yogi and the Commissar*, MacMillan, London 1945.

return to his characteristic philosophical motifs: the emphasis on contingency, on the ambiguity and opacity of history, on the absence of any "objective" or "scientific" parameter to which one could appeal in choosing or judging – in short, on the hazardous and always contradictory character of human existence.[39]

Thus returns the distinctive configuration of Merleau-Ponty's reflection: to render possible a discourse on that objectivity (the world, nature, history) that has been dismissed as an initial foundation, he attempts to ground – within subjectivity itself and within its constitutive structures – the very presuppositions of a necessary opening to the world. Yet all of this now begins to appear in tension with the transformed political landscape of the 1950s. Sartre himself, in an editorial of 1961 published in *Les Temps Modernes*[40], identifies precisely those years as the moment when Merleau-Ponty encountered the difficulty of continuing along the theoretical path traced in *Humanisme et terreur:* "When everything exploded, around 1950, he saw nothing but wreckage." It is here, then, that the problematization of the role of Marxism begins – a process that Merleau-Ponty will articulate, throughout the years around 1950, in a series of articles published in *Les Temps Modernes* and *L'Express*, culminating in the synthetic statement offered in his 1955 work, *Les Aventures de la dialectique (The Adventures of the Dialectic).*

The confrontation with the political events of that period – the publication of Trotsky's alleged testament, the revelations concerning the existence of forced labor camps in the USSR, and finally the outbreak of the Korean War – had provoked uncertainty, crisis, and disorientation among the ranks of the Left. In a climate so deeply different from that of the Resistance and from the brief moment of renewed interest in

39 See Merleau-Ponty, M. *Humanism and Terror.* quot., p. 105.

40 Cf. Sartre, J.-P., *Merleau-Ponty vivant*, "Les Temps Modernes", 184-185, 1961, pp. 304-376.

Marxism within the MRP, Merleau-Ponty asks: in the face of the disconcerting events of Stalin's Russia, can one still truly believe that Marxism is capable of guaranteeing humanity "a more human future"? And if it were not true that communism prepares an era of freedom, and if it were instead that very negation of the human that its critics see in it? Already in the pages of *Humanism and Terror* the first doubts had surfaced, accompanied by evident reconsiderations – tangible signs of a possible, perhaps inevitable, change of direction: from the Marxist attentisme of 1947, a position that in a certain sense justified communist violence, to a decisive non-communism, a position that would remain unchanged until the philosopher's death. Il problema di *Umanismo e terrore* ritorna quindi, insistente e ineludibile, e si approfondisce di volta in volta negli articoli apparsi su *Les Temps Modernes* a partire dal 1948: "Communisme et anti-communisme"[41]; "Marxisme et superstition"[42]; "Les jours de notre vie"[43]. Here, in fact, the need to take distance from an overly optimistic and confident adherence to the validity of the Soviet experience becomes increasingly urgent. The judgment on that experience grows progressively harsher. Yet this position – certainly painful and long meditated – did not entail any adherence to the *a priori* anti-communism that characterized so much of French culture at the time; rather, it arose precisely from the refusal to accept the simplistic and polemical identification of communism with fascism. But it was not only a matter of this refusal. The philosopher's decision also entailed the necessity of an increasingly clear and critical distancing that concerned not merely communist politics in the strict sense, but also the theoretical positions of Marxism itself.

41 Merleau-Ponty, M., *Communisme et anti-communisme*, "Les Temps Modernes", 34, 1948, pp. 175-188.

42 Id., *Marxism and superstition*, in *Signs*, pp. 261-262.

43 Id. (with J.-P. Sartre), *Les jours de notre vie*, "Les Temps Modernes", 51, 1950, pp. 1153-1168.

Merleau-Ponty now pays the price for his initial humanistic interpretation of Marxism. For the French philosopher, it becomes evident that the Stalinist *betrayal* of the revolution consequently entails a crisis of faith in socialism and in the theoretical credibility of Marxism: if Stalinism does not realize the human, but rather turns against the human, this demonstrates that even at the theoretical level Marxism does not in fact place the human at the center of history.

Adventures of the Dialectic gathers together the threads of this progressive shift – a shift that, beginning from the recognition of the crisis of unity among the Left and from reflection on international events, led Merleau-Ponty to become increasingly convinced of the impossibility of continuing the mediating function between the communist world and left-wing intellectuals, a function he had himself assumed in the immediate postwar period.

We must say that, if on the one hand the French thinker had, earlier than others, attempted to influence the evolution of Marxism by working from within, maintaining a permanent role of critique and of stimulus, on the other hand it is equally true that, in the years of the Cold War, he quickly distanced himself from Marxism and from communism. For one who had tried to discern in the USSR the place where a socialism understood as the full realization of the human could be constructed, the situation of those years appeared to impose a radical alternative: either the USSR and the communist parties did not, in reality, aspire to socialism, or socialism itself did not realize – but negated – that humanism whose echo the early writings of Marx had seemed, for a moment, to reawaken within Marxist thought.

It is in this second direction that Merleau-Ponty's reflection now proceeds: toward a problematization of the political role of Marxism[44] that places at its very center the theme of dialectic,

44 Moravia, S. in his article entitled *La crisi della generazione sartriana* (cited above), observes—regarding Merleau-Ponty's

the foundational principle in Marx of both philosophy and praxis.

If *Humanism and Terror* had emphasized the youthful dialectic of Marx as the guarantor of the role of subjectivity in history – a safeguard against the always possible regressions into scientistic frameworks – it now appears to Merleau-Ponty that Marxism has come to be entirely characterized as a dogmatic objectivism. For to consider the revolutionary task as something objective, beyond and independent of the subjective consciousness that the proletariat may have of it, constitutes, for the philosopher, the irreversible break of the interrelation between subjectivity and objectivity and the introduction into the historical process of a vantage point other than that of concrete human beings.

The critique of the Marxist concept of objectivity is linked – not by chance, in the 1955 work – to the meditation on the trajectory of "Western Marxism," and above all on that Lukács whose early theses had appeared to Merleau-Ponty to express the identity of subject and object within the historical process, and on which, to such a large extent, he had constructed his '*marxism d'attente*'. Yet the orthodoxy of the 1940s and 1950s, and the repudiation of the young Lukács himself, seemed to demonstrate the impossibility of sustaining such an adherence.

Merleau-Ponty then returns to discover in Marx himself the negation of the humanism that had originally oriented his reflection: "The conflict between "Western Marxism" and Leninism is already found in Marx as a conflict between dialectical thought and naturalism, and the Leninist orthodoxy eliminated Lukacs' attempt as Marx himself had eliminated

political shift as articulated in the 1955 work – that: "What remains is the fact that this is an examination of conscience carried out before the deluge: a remarkable capacity to discern the authentic fundamental problems and to take on them a clear and stimulating position, open to new times" (p. 439, my translation). See also pp. 441 and following.

his own first "philosophical period"[45]. The misunderstanding of later Marxism, then, derived directly from the Marx who, in the pursuit of an illusory scientificity, abandoned his early theses. The adoption of the concept of objectivity, the search for a scientific point of view on history, thus appear as the sign of a renunciation: a renunciation of the moment of subjectivity, of élan, of revolutionary impulse. When, after 1850, *"philosophical Marxism"* is replaced by *"scientific socialism"*, everything changes. Here lies the decisive point of rupture; it is by virtue of that shift that Marxism becomes

> is a fast and loose philosophical system which disarms the dialectic by denying to the subject the judgment of history and the intrinsic appreciation of literature and politics; but it is a system which leads one to believe that the dialectic continues to function in the infrastructures and in the mysterious future which the infrastructures are preparing; and it is a system which honors the dialectic from afar and, without practicing or disavowing it, annuls it as a critical instrument, keeping it only as point of honor, justification, and ideology.[46]

At the origin of the present-day degeneration of Marxism lies, therefore, the renunciation – on the part of the mature Marx – of philosophy, of dialectic, of subjectivity, and the choice in favor of a supposed "science of history" which abstracts from the constitutive presence of the human being in order to seek in things the "principle" of historical processes. This, then, is the final face of Marx: he "continually increased the weight of the objective factors of history". In fact, "the beautiful parallelism in the young Marx between the realization of philosophy and the realization of socialism was destroyed by 'scientific socialism' to the benefit of the infrastructures"[47]. It is striking to perceive how, between the lines of this bitter denunciation, Merleau-Ponty reveals

45 Merleau-Ponty, M., *Adventures of the Dialectic,* quot., p. 64.
46 Therein, p.72.
47 Therein, p.84.

not only a categorical critique of Marxism, but above all the courage to go against his own former position. From the initial accusation directed at the USSR – held responsible for having betrayed the revolution – Merleau-Ponty moves to the thesis of the *inevitability* of bureaucratic-dogmatic outcomes.

This means that, if it is not possible to realize within the revolutionary process a genuine balance between objectivity and subjectivity, then history truly possesses a fundamental opacity, and human beings will never succeed in remaking it in their own image. The philosopher therefore shifts from a stance of critical support of the PCF to the theorization of the *Third Force*, from which arises his polemic against the contemporary – and opposite – evolution of Sartre[48]. The critique of Marxist dialectics is thus linked to that very philosophy of subjectivity through which Merleau-Ponty had once attempted to find a point of convergence with Marxism. The difficult balance between subjectivity and objectivity, which *Humanism and Terror* had sought to maintain by elaborating a "Marxist attentisme" will take shape in the subsequent evolution of Merleau-Ponty's thought, within the horizon of a research increasingly oriented toward ontology. Moreover, Merleau-Ponty's reflection on Marxism cannot be reduced to a mere philosophical-political *engagement*: it is grounded in an undeniable and complex theoretical and

48 Further on this debate, see: *Sartre/Merleau-Ponty, Les lettres d'une rupture*, in "Magazine Littéraire", cit.; Italian transl. by Calabrò D., and Lisciani-Petrini, E., *Sartre/Merleau-Ponty. Il carteggio della rottura*, quot., pp. 193–221, and later Italian transl. and edited by E. Lisciani-Petrini, *Sartre/Merleau-Ponty. Le lettere della rottura*, quot., pp. 21–56; Calabrò, D., *The Correspondance. Sartre and Merleau-Ponty*, Mimesis International, London–Milan, 2018; Lisciani-Petrini, E., *Merleau-Ponty–Sartre: una insanabile divergenza filosofico-politica*, quot., pp. 61–90; Kirchmair, R., *Alle radici di un equivoco. Filosofia e politica in Sartre e Merleau-Ponty*, in "aut aut", 381, 2019, pp. 91–119; Caeymaex, F., Cormann, C., *Sartre/Merleau-Ponty, andate e ritorni*, in "aut aut", 381, 2019, pp. 120–148.

ontological task.[49] It is within this practice of *thinking* being-in-the-world and of *inhabiting* its sense that one must understand the *"what is to be done"* of Merleau-Ponty's philosophy, as well as its specific *humanitas*, both on the specifically political level and – inseparably connected to this – on the level of intersubjective relations and the formative processes of individuals. A *humanitas* that can therefore be defined as *impersonal*, as Enrica Lisciani-Petrini finely observes in the volume *Vita quotidiana. Dall'esperienza artistica al pensiero in atto,* reinterpreting Merleau-Ponty: "around our personal experience there appears a margin of existence that is almost impersonal, which, so to speak, goes on of itself, to which I entrust myself in order to sustain my life … and to which one must first of all belong; but—with accents that echo Bergson—he comes to say that thought 'is nothing interior, does not exist outside the world, but is formed there, in the world, outside each of us.' Like a 'vortex' […] that unfolds within reality and in which we are implicated. As though we were 'rays' in a specular play that 'radiates around an entirely virtual center.' To the point that—and here one sees the radical consequence of this discourse which, from afar, takes its cue from Freud—'there is no identity, nor non-identity,' but 'there is an inside and an outside turning around one another.' Thus, 'I', truly, is no one; it is the anonymous."[50] This nameless

49 See VV.AA., *Merleau-Ponty aux frontières de l'invisible*, Mimesis, Milan 2003; Barbaras, R., *De l'être du phénomène. Sur l'ontologie de Merleau-Ponty*, J. Millon, Grenoble 1991, and Id., *Le tournant de l'expérience. Recherches sur la philosophie de Merleau-Ponty*, Vrin, Paris 1998; Carbone, M., *The Flesh of Images: Merleau-Ponty between Painting and Cinema*, SUNY Press, New York 2016; Lanfredini, R., Vitale, S., *Pensare in ultimo. Merleau-Ponty dopo Merleau-Ponty*, Moretti & Vitali, Bergamo 2014; Lisciani-Petrini, E., *La passione del mondo. Saggio su Merleau-Ponty*, quot.; Vanzago, L., *The Voice of No One. Merleau-Ponty on Nature and Time*, Mimesis International, London–Milan 2017.

50 Lisciani-Petrini, E. *Vita quotidiana. Dall'esperienza artistica al pensiero in atto*, Bollati Boringhieri, Turin 2015, pp. 53-54 (my

"I" tells the opaque story of its institution, of its power, of its activity/passivity[51].

Thus, to conclude, in response to Sartre's invective in the Carteggio of 1953, where the latter had declared the impossibility of adopting a political stance – "today the philosopher cannot adopt a political position" – Merleau-Ponty replies by writing: "In reality, we do not live in two separate orders, that of consciousness and that of the event-as-thing. There is no philosophy that is detached from the plane of events. Most action takes place neither in things nor in spirits, but rather in the between that stretches between events and pure thoughts, in the thick stratum of symbolic choices, which operate more through their meaning than through their efficacy."[52] Certainly, the passages placed in the epigraph of this essay belong within the horizon of an *engagée* and militant reflection, which – as such – must always be put to the test, continuously measured against the unfolding of those concrete historical-political events in which it is immersed. It is in this direction that one must understand Merleau-Ponty's phenomenological claim, never abandoned, namely, the originary co-belonging of consciousness and world.

This *ante-predicative sphere*, already investigated by Merleau-Ponty in his 1945 work, indicates that consciousness is not an absolute beginning, but rather the continuous resumption of a movement that has always already begun within lived experience. On the plane of political action, there does not exist solely the binary relation between "human being–historical

---

translation).

51 Id., *Attività/passività: l'invisibile di Merleau Ponty*, "Chiasmi international", 1, 1999, pp. 169-185. See also: Lapierre C., *Entre conscience et inconscient. Une lecture du problème de la passivité chez Sartre et Merleau-Ponty,* in "Chiasmi International", n. 17, 2015, pp. 95-132

52 Carteggio *Sartre/Merleau-Ponty. Le lettere della rottura*, quot., p. 51 (my translation). See also: *The Correspondance: Jean-Paul Sartre and Maurice Merleau-Ponty,* quot.

event," separated by an abyss of nothingness that could be overcome only through a pure act; rather, there also exists the symbolic space of history as continuous reinterpretation, mediation, and divergence.

For Merleau-Ponty, to engage oneself does not mean plunging into the dramatic alternatives of history in order to identify oneself entirely with one of them, for this would strip *engagement* of its very peculiarity – its being, that is, a continuous re-projection, an unceasing re-elaboration of the meaning of events – and would condemn it to silence, to self-censorship (silence and self-censorship which, according to Sartre, Merleau-Ponty should have imposed upon himself within the journal *Les Temps modernes*). To engage is, rather, above all, to situate oneself within historical concreteness, while nonetheless maintaining between oneself and events a *vigilant distance* which is nothing other than the very recognition of that veil of ambiguity and uncertainty that separates us from them.

At the conclusion of his long and decisive critique of Sartre, Merleau-Ponty forcefully gathers these ideas in a passage of *Adventures of the Dialectic*, in which one senses the weight of all the lived experiences and disappointments – and at the same time the firm intention to go further:

> The question is to know whether, as Sartre says, there are only men and things or whether there is also the interworld, which we call history, symbolism, truth-to-be-made. If one sticks to the dichotomy, men, as the place where all meaning arises, are condemned to an incredible tension. Each man, in literature as well as in politics, must assume all that happens instant by instant to all others; he must be immediately universal. If, on the contrary, one acknowledges a mediation of personal relationships through the world of human symbols, it is true that one renounces being instantly justified in the eyes of everyone and holding oneself responsible for all that is done at each moment. But since consciousness cannot in practice maintain its pretension of being God, since it is inevitably led to delegate responsibility – it is one abdication for another, and we prefer the one which leaves

> consciousness the means of knowing what it is doing. To feel responsible for everything in the eyes of everyone and present to all situations – if this leads to approving an action which, like any action, refuses to acknowledge these principles, then one must confess that one is imprisoned in words. If, on the contrary, one agrees that no action assumes as its own all that happens, that it does not reach the event itself, that all actions, even war, are always symbolic actions and count as much upon the effect they will have as a meaningful gesture and as the mark of an intention as upon the direct results of the event-if one thus renounces 'pure action,' which is a myth (and a myth of the spectator consciousness), perhaps it is then that one has the best chance of changing the world.[53]

We must undoubtedly recognize that the perspective advanced by Merleau-Ponty – grounded in a culture of liberation and commitment and centered on the autonomy of the intellectual – constituted not only a valid model of *engagement*, but at the same time offered a decisive return to the living sources of philosophy. This occurred according to a certain *"style"* which, beginning from the critique of the inexhaustible dichotomies of subject/object, theory/praxis, idea/action, and developing as a continuous interpretation of interwoven paths (phenomenology, existentialism, Marxism, structuralism, ontology), situates itself not *beyond*, but *within* that *Lebenswelt* which, far from being the site of the unfolding of an Absolute History, is rather the place of "shadows", "displacements", and "metaphors".

Thus, an *intentional History*, an expression of the implicit, the unthought, the *shadow* that always accompanies the philosopher. Reflection therefore does not become systematic and definitive inquiry aimed at total transparency; rather, it becomes continuous meditation, a perennial question of meaning – of that meaning which History offers to us yet never makes explicit. In proposing a research that is indeterminate

53 Merleau-Ponty, M., *Adventures of the Dialectic*, quot., pp. 200-201.

and never completed, in affirming the certainty of a history always to be reinterpreted and of a freedom always to be reconquered, we rediscover the eminently political task and role of Merleau-Ponty's thought. It is therefore quite evident how far removed this is from Sartre's perspective, which "ignores density, depth, the plurality of planes, the background worlds".[54]

And it is here, in our view, that one must place Merleau-Ponty's decisive response to Sartre in the 1953 Correspondence: namely, the affirmation of a directly political role for philosophy, a role that prevents any absolutization, any clear and distinct vision, any *survol* of thought and thus any *hybris* of identity. This is, without doubt, the theme of Merleau-Ponty's *classicism* – the same classicism that he attributed to Marx[55], – of his continuing weight upon our present as a valid interlocutor for our political practice, for our very idea of philosophy, and for the *entrelacs* between the two. It is therefore necessary that the fundamental themes of Merleau-

54 See Id., *The Visible and the Invisible,* quot., p.68

55 In the Introduction to *Signs*, Merleau-Ponty expresses his idea of Marx's classicism as follows: "We are saying that a re-examination of Marx would be a meditation upon a classic, and that it could not possibly terminate in a nihil obstat or a listing on the Index.[…] The history of thought does not summarily pronounce: This is true; that is false. Like all history, it has its veiled decisions. It dis mantles or embalms certain doctrines, changing them into "messages" or museum pieces. There are others, on the contrary, which it keeps active. These do not endure because there is some miraculous adequation or correspondence between them and an invariable "reality" – such an exact and fleshless truth is neither sufficient nor necessary for the greatness of a doctrine – but because, as obligatory steps for those who want to go further, they retain an expressive power which exceeds their statements and propositions. These doctrines are the classics. They are recognizable by the fact that no one takes them literally, and yet new facts are never absolutely outside their province but call forth new echoes from them and reveal new lustres in them"; the quotation can be found on pp. 10–11.

Ponty's reflection be, above all today, classically present, so that the "new" in contemporary culture and politics may be confirmed in a forward-looking orientation that is always a critical and coherent reprise of past History.

> We are not spectators of a closed history; we are actors in an open history, our praxis introduces the element of construction rather than knowledge as an ingredient of the world, making the world not simply an object of contemplation but something to be transformed. What we cannot imagine is a consciousness without a future and a history with an end. Thus, as long as there are men, the future will be open and there will only be a probabilistic calculation and no absolute knowledge. Consequently 'the dictatorship of the truth' will always be the dictatorship of a group.[56]

The danger of conceiving truth as a structure of domination remains the central theme that traverses Merleau-Ponty's entire philosophical reflection, shaping his intellectual experience during one of the most arduous and turbulent phases of postwar European history. It is an experience that undoubtedly retains long-term relevance, a certain "classical" dignity of thought that refers to a perhaps forgotten culture of free philosophical and political *engagement*, which we should attempt, today, to recover for ourselves.

As a practice of mediation, questioning, and responsibility, political action, in this perspective, is always situated and contingent, ambiguous and structurally opaque; it is inscribed in the *between*, in the intermediate thickness that links thought to event, consciousness to reality, the subject to the other. This is, as suggested at the outset, a militant philosophy in the highest sense of the term: a way of thinking that does not seek to soar above history, but instead exposes itself to risk, dwelling in ambiguity as the constitutive condition of human action. Such a conception, far from being confined to the European postwar context, proves to be particularly relevant

56 Id., *Humanism and Terror*, quot., p. 92.

today. In an era marked by the crisis of grand narratives, the fragmentation of social bonds, and the progressive hollowing out of democratic institutions, Merleau-Ponty offers us a vision of politics as the *institution of sense*, as an open dialogical praxis that resists both ideological closure and postmodern nihilism. His reflection invites us to recognize the political *sense* implicit in every form of perceptual and linguistic experience, showing that freedom does not consist in the absolute autonomy of the subject, but in the ability to respond – always provisionally and openly – to the weave of relations and events that constitute us.

In this direction, Merleau-Ponty's lesson re-emerges as one of the most fertile theoretical resources for rethinking contemporary politics: not as a technique of power or a management of emergencies, but as an ethics of coexistence, as a praxis of sense that keeps the horizon of history open and counters every form of "dictatorship of truth." His conception of the political is thus that of a site of an ontology of relation, in which freedom and truth are not imposed from above, but are instituted each time in the fragile and concrete encounter between subjects and the world[57].

It is in this permanent openness, in this thinking of history as a field of possibilities and sense, that the profound contemporaneity of his philosophy is revealed.

57 Di Martino C., *The body of freedom,* in "Chiasmi International", n. 11, 2009, pp. 419-430.

PART THREE
A DIASTOLE WITHOUT SYSTOLE

CHAPTER 1
AN OPEN DIALECTIC

§1. *The comparison with Hegel*

The analysis of Merleau-Ponty's "phenomenological" itinerary concludes with a questioning of the meaning of Being, which, by definition, must remain unfinished. Yet this point of arrival, that is, the theoretical gain implicit in the revision of Husserlian phenomenology, becomes – in turn – a new and more controversial beginning. For Merleau-Ponty, it involves returning to those very core ideas that had characterized his philosophical reflection up to that moment and putting into question not only their validity but also, and above all, the entire unity of sense. What operation did he intend to carry out? What external stimuli influenced his speculative itinerary? For what theoretical necessity did the French philosopher consider it indispensable to engage with Hegel of the *Phenomenology of Spirit*?

Since the work of 1945, Merleau-Ponty found himself in a certainly not easy position. Indeed, overcoming the paradoxes of Husserlian phenomenology required not only a genetic radicalization of the concept of *Lebenswelt*, but also, and above all, the recognition of the idea of dialectic in Hegel, thus realizing the profound and secret convergence of the two phenomenological matrices. The operation he was therefore preparing to undertake was to allow Hegelian dialectic of the *Phenomenology of Spirit* to penetrate Husserlian phenomenology, with the intent of finding in Hegel the possibility of a decisive renewal of phenomenology – one that would ground the latter not only on the phenomenological description of things themselves but also on the dialectical identity of knowledge and the Absolute.

Moreover, the reason for the convergence between the two phenomenologies was highlighted, in those same years, by Kojève, who – in the work entitled *Introduction to the Reading of Hegel* – wrote:

> The attitude of the philosopher or the 'scientist' (= the Wise Man) with respect to Being and to the Real is one of purely passive contemplation, and that philosophic or "scientific" activity reduces to a pure and simple description of the Real and of Being. The Hegelian method, therefore, is not at all 'dialectical': it is purely contemplative and descriptive, or better, phenomenological in Husserl's sense of the term. In the Preface and the Introduction to the Phenomenology, Hegel insists at length on the passive, contemplative, and descriptive character of the 'scientific' method. He underlines that there is a dialectic of 'scientific' thought only because there is a dialectic of the Being which that thought reveals. As soon as the revealing description is correct, it can be said that *ordo et connexio idearum idem est ac ordo et connexio rerum*; for the order and the connection of the real are, according to Hegel, dialectical.[1]

Thus, the French scholar concludes shortly thereafter, in a peremptory statement that returns to Hegel's non-dialectical method: "what can be said has already been said by Edmund Husserl about his own 'phenomenological' method, which he quite wrongly opposed to the Hegelian method with which he was not familiar. For in fact Hegel's method is nothing but the method that we nowadays call 'phenomenological.'"[2].

Certainly, the influence of this text by Kojève was, in many respects, decisive for the continuation of Merleau-Ponty's path

1 Kojève, A., *Introduction to the Reading of Hegel*, Cornell University Press, N.Y. 1980, pp.171-172.

2 Therein, p. 195.

in the direction of ontology. For this very reason, it is necessary to situate the French philosopher's discovery of Hegel within a broader and more detailed historical framework. We refer, in particular, to the revival, beginning in the late 1920s, of Hegelian studies within the existentialist milieu.

Jean Wahl, in 1929, with *Le Malheur de la conscience dans la philosophie de Hegel*, had undoubtedly marked a turning point in Hegelian studies; and this occurred not only within French culture. "Hegel's philosophy – Wahl wrote – cannot be reduced to a few logical formulas. Or rather, such formulas conceal something that is not of purely logical origin. Dialectic, before being a method, is an experience through which a spirit continually proceeds beyond what it *is*."[3] The systematic Hegel of the *Encyclopedia* and the *Logic* was thus abandoned in favor of the concrete and problematic Hegel of the *Phenomenology*. "*Phenomenology* – Wahl reiterated – [...] no longer appears solely as an introduction to the doctrine, but also as a point of arrival."[4]. In essence, for Wahl, the Hegelian system, in which concepts appear marvelously manipulated and ordered, is the expression of a living experience, a response to a problem that is not merely intellectual. Wahl's work, moreover, was situated within an already extensive panorama of Hegelian studies: in 1930, Gurvitch *Les tendances actuelles de la philosophie allemande*[5]. The following year, Hartmann published *Hegel et le problème de la dialectique du réel*[6] in the authoritative *Révue de métaphysique et de morale*, in which it was asserted that the problematic richness of Hegel did not lie so much in the foundations of his system or in the most abstractly speculative constructions, but rather in the

3 Wahl, J., *Le Malheur de la conscience dans la philosophie de Hegel*, PUF, Paris 1951, pp. 1-2 (my translation).

4 Ibid.

5 Gurvitch, G. *Les tendances actuelles de la philosophie allemande*, Vrin, Paris 1930.

6 Hartmann, N. *Hegel et le problème de la dialectique du réel*, "Revue de Metaphisique et de Morale", Juillet-Septembre 1931.

investigation of the various fields of concrete spiritual life, in which his power of insight proved incomparable. In this sense, Hartmann considered Hegel's fundamental work to be the *Phenomenology*, rather than the *Logic* or the *Encyclopedia*, and explained that Hegel's philosophy, regarded in its most valuable contribution, was an immense and unique phenomenology of spiritual struggle, of the effort undertaken by man to create, understand, and organize his own world. In the years 1932-'33, Koyré and Kojève contributed decisively to the rediscovery of Hegel through their university courses. Within the debate on Marxism, Lefebvre, in turn, introduced and elaborated in 1939, with *Dialectical Materialism*[7] the problem of the validity and significance of Hegelian idealism, attempting a reconstruction of Marx's critical examination of Hegel's philosophy conducted between 1843 and 1859, in the conviction that "Hegel's ambition remains valid and coincides with that of philosophy" and then "Perhaps it is possible to transcend Hegelianism on its own terms, from inside, by starting from its own contradictions and preserving what is essential in its mode of operation."[8] Ultimately, the debate on the *Phenomenology of Spirit* – to the knowledge and dissemination of which the first translation by Jean Hyppolite, carried out between 1939 and '41, had contributed decisively, and who would later, in 1946, publish the famous reconstructive essay *Genesis and Structure of Hegel's Phenomenology of Spirit* – was in full swing in France during the 1930s and '40s. It is precisely in this period that Merleau-Ponty's reflection on Hegel takes place, within which Marx and Husserl are also present. These are, therefore, three thinkers placed in a fruitful relationship of dialectical complementarity, in an effort of reflection that simultaneously seeks to distinguish them and to resolve them into a new philosophical perspective.

---

7 Lefebvre, H., *Dialectical Materialism*, University of Minnesota Press, Minneapolis 1968.

8 Therein, p. 80.

The occasion for a clarifying intervention within the existentialist debate on Hegel was offered to Merleau-Ponty by a lecture held by Hyppolite[9], one of the first to have initiated the translation of existentialism into Hegelian terms. In the essay *Hegel's Existentialism*, published in April 1946 in the journal *Les Temps Modernes* and later included in the volume *Sense and Non-Sense*, Merleau-Ponty explicitly declares:

> All the great philosophical ideas of the past century – the philosophies of Marx and Nietzsche, phenomenology, German existentialism, and psychoanalysis – had their beginnings in Hegel; it was he who started the attempt to explore the irrational and integrate it into an expanded reason which remains the task of our century. He is the inventor of that Reason, broader than the understanding, which can respect the variety and singularity of individual consciousnesses, civilizations, ways of thinking, and historical contingency but which nevertheless does not give up the attempt to master them in order to guide them to their own truth. But, as it turns out, Hegel's successors have placed more emphasis on what they reject of his heritage than on what they owe to him. If we do not despair of a truth above and beyond divergent points of view, if we remain dedicated to a new classicism, an organic civilization, while maintaining the sharpest sense of subjectivity, then no task in the cultural order is more urgent than re-establishing the connection between on the one hand, the thankless doctrines which try to forget their Hegelian origin and, on the other, that origin itself. That is where their common language can be found and a decisive confrontation can take place. Not that Hegel himself offers the truth we are seeking [...] but all our antitheses can be found in that single life and work. There would be no paradox involved in saying that interpreting Hegel means taking a stand on all the philosophical, political, and religious problems of our century.[10]

---

9 See J. Hyppolite, *L'existence dans la "Phénoménologie" de Hegel*, conference held at the Société des Etudes germaniques on February 16, 1946 and published in the "Revue des Etudes germaniques", n° 2, avril-juin 1946; later in *Figures de la pensée philosophique. Ecrits de Jean Hyppolite (1931-1968)*, P.U.F., Paris 1971, pp. 92-103.

10 Merleau-Ponty, M., *Sense and non-sense*, quot., pp. 63-64.

The dialogue with Hegel is thus initiated: for Merleau-Ponty, it becomes necessary to "always anew" begin speaking of phenomenology but now starting from the theoretical implications present in the *Phenomenology of Spirit*. The Hegel of 1807, in fact, offers the French philosopher new tools to free existentialism from the impasses of subjectivism, and Marxism from the economists tendencies from which it had not yet disentangled itself, partly due to certain directions in the Marx of the 1850s–'60s. But it is not only this: through Hegel, it becomes possible for Merleau-Ponty to bring existentialism and Marxism into encounter and to refound both on phenomenological grounds. His preference for the non-systematic Hegel is grounded primarily in the pursuit of the humanistic aspects of Hegelian reflection, where it is possible to discern new and compelling developments regarding the conception of the individual and his internal contradiction, beginning with the examination of the negative, of death, and of the dialectical process. Thus, the French philosopher does not fail to note that in Hegel conflicts are not resolved, but reemerge and are renewed beyond any possible reconciliation, to the point that duality does not remain external to the Absolute but constitutes its intrinsic principle. In this sense, it is extremely significant that Merleau-Ponty aligns with the theoretical lines of Hyppolite, for whom "the determination of the foundation of the historical fact coincides without doubt with the determination of the foundation of the human fact." Consequently, man is placed "as the criterion and principle of the historical world, as an element of explication and coordination of the context of historical facts in their complex givenness; intentional analysis provides this thematization with its specific method."[11] This is an analysis in many respects

11 Hyppolite, J., *Genesis and Structure of Hegel's Phenomenology of Spirit*, Transl. by S. Cherniak and J. Heckman, Northwestern University Press, Evanston, Illinois 1974, pp. XII-XIII (my translation).

close to the Marxist interpretation of Hegelian thought, aimed therefore at emphasizing the necessity of reading Marx starting from the *Phenomenology of Spirit*. Indeed, in those years, the revival of Hegelian studies in France was integrated and mediated with the renewed interest in Marx's early writings – from *A Contribution to the Critique of Hegel's Philosophy of Right* to the *Economic and Philosophical Manuscripts of 1844*, and to the *Theses on Feuerbach*. It is in this direction that Merleau-Ponty moves, seeking to unite – with a view to a new conception of history – through Hyppolite, Hegel's humanism, and, through Lukács, the humanism of the young Marx.

The complete polarization of research on the systematic apparatus of Hegelian doctrine had, before the 1930s, prevented attention from being directed toward the extraordinarily rich context of observations Hegel carried out within the framework of man's historical and social reality. It is precisely from these observations that Merleau-Ponty begins anew, certainly stimulated by the innovative currents of contemporary thought – such as existentialism, phenomenology, and Marxism – but motivated above all by the idea of a philosophical project that would take him beyond his early works, based primarily on the total disintegration of the systematicity of Hegelian doctrine and, secondly, on the expansion of Kantian-Husserlian phenomenological investigation toward the dynamic conception of dialectic.

With this, the myth of the all-encompassing Absolute disappears, giving way to a dialectical movement which, far from being defined as an instrument for the annulment of the finite in the infinite, becomes a heuristic principle, a key suitable for revealing the figure of the chiasm, of reciprocal envelopment, of the *"entrelacs"*. The developments and deepening of this project would also lead Merleau-Ponty to a clearer awareness of the necessarily finite and unfinished character of his own inquiry, and to the definitive abandonment of any rational pretension to constitute the world in favor of a truly open, dialogical dimension of human investigation.

Merleau-Ponty's procedure is therefore aimed at dissolving the most typically theological and metaphysical aspects of Hegelian doctrine in favor of all those contributions that could usefully be brought closer to the philosophical reflection of his time. In this framework, one of the Hegelian contributions that appeared fundamental to the French philosopher was undoubtedly the interpretation of history. Merleau-Ponty's analysis on this point is perfectly clear: he intends to demonstrate the validity of Hegel's 1807 conception of history at the expense of the alleged resolution of the internal contradictions of history itself offered by Hegel in 1827. This is what Merleau-Ponty declares without reservation in an extensive passage of *Hegel's Existentialism:*

> Kierkegaard, the first to use "existence" in the modern sense of the word, deliberately set himself up in opposition to Hegel. The Hegel he had in mind was the late Hegel, who treated history as the visible development of a logical system, who sought in the relationships between ideas the final explanation of events, and who subordinated the individual experience of life to the life appropriate to ideas, as to a destiny. This Hegel of 1827 offers us nothing but a "palace of ideas," to use Kierkegaard's phrase, where all historical antitheses are overcome, but only by thought. Kierkegaard is right in objecting that mere thought is not enough to enable the individual to overcome the contradictions facing him, that he is faced with dilemmas neither term of which he can accept. This last Hegel has understood everything except his own historical situation; he has taken everything into account except his own existence, and the synthesis he offers is no true synthesis precisely because it pretends ignorance of being the product of a certain individual and a certain time. Kierkegaard's objection, which is in profound agreement with that of Marx, consists in reminding the philosopher of his own inherence in history [...] But if the Hegel of 1827 may be criticized for his idealism, the same cannot be said of the Hegel of 1807. The *Phénoménologie de l'Esprit* is a history not only of ideas but of al the areas which

> reveal the mind at work: customs, economic structures, and legal institutions as well as works of philosophy.[...] The real debate between Marx and Hegel has nothing to do with the relationship of ideas to history; rather, it involves the conception of historical movement, which ends for the Hegel of 1827 in a hierarchical society whose meaning is accessible to none except the philosopher but Which the Hegel of 1807 perhaps saw culminating in a genuine reconciliation between men. What is certain in any case is that the *Phénoménologie de l'Esprit* does not try to fit al history into a framework of pre-established logic but attempts to bring each doctrine and each era back to life and to let itself be guided-by their internal logic with such impartiality that all concern with system seems forgotten.[12]

The protest of Kierkegaard and Marx, who seek to return philosophy to its historical immanence – entirely valid against the Hegel of the *Logic* – is certainly, in Merleau-Ponty's view, not valid against the Hegel of the *Phenomenology*, which attempts to "explore the irrational and integrate it into an expanded reason"[13]. In *Phenomenology of Spirit*, man is not immediately absolute consciousness, but finite, fractured consciousness – unhappy consciousness, anxiety: a life given to itself that seeks to understand itself, to transcend every determination, to universalize itself. Against the alternative of the objective sense or the radical nonsense of history, Merleau-Ponty suggests that existence itself, as adherence and immanence to the world, is already the having-and-giving of meaning. And this meaning to which we are destined is, from the very beginning, already communication; it is therefore not possible to speak of the solipsism of consciousness. If the existence of the other is a scandal for consciousness, it nonetheless ensures that there is an initial meaning, because through his very existence, man opens himself to meaning, to communication with the other. I have always already been grafted onto others.

12 Merleau-Ponty, *Sense and Non-sense*, quot., pp. 64-65.
13 Therein, p. 63.

Being-in-the-world is synonymous with a meaning that ignites, with a fabric that begins to weave itself without presupposing a pre-given pattern of essences, or without my having to consider myself as the sole weaver of the entire fabric. All this reveals that our relation to the social is deeper than any expressed perception or judgment, and reconnects to the problems of the body itself, of the natural world, of birth and death, forming a single, unified problem with them. Merleau-Ponty entertains the idea of a single history of expression, stemming from that decisive passage in Hegel (the transition from the Absolute as substance to the Absolute as subject), which, for Heidegger, is the apex of the trajectory inaugurated by Descartes – namely, the identification of *subjectum* and *res cogitans* – but for the French philosopher constitutes the advent of a new conception of the Absolute: a dialectical Absolute, fluidified within experience, severing ties with the Cartesian-Kantian perspective of the modern age, and presenting itself as a new beginning of philosophizing.

In this light, Hegelian dialectic appears capable of allowing us to understand the intricate entanglement of man and world, a key suitable to unlock the secrets of an existence devoted to restlessness and therefore never fully pacified – a moment of verification of the totality and dynamism of the real, of the problematic nature of existence, and of history itself. And so, thanks to Hegel, Merleau-Ponty conclude:

> The "subject" is no longer just the epistemological subject but is the human subject who, by means of a continual dialectic, thinks in terms of his situation, forms his categories in contact with his experience, and modifies this situation and this experience by the meaning he discovers in them. In particular, this subject is no longer alone, is no longer consciousness in general or pure being for itself. He is in the midst of other consciousnesses which likewise have a situation; he is for others, and because of this he undergoes an objectivation and becomes generic subject. For the first time since Hegel, militant philosophy is reflecting

> not on subjectivity but on intersubjectivity. Transcendental subjectivity, Husserl pointed out, is intersubjectivity.[14]

In these final lines, we can see how capable Merleau-Ponty was of establishing a fruitful interrelation between Hegelian dialectic and Husserlian phenomenology, gradually isolating precise points of convergence between the Hegelian themes of the *Phenomenology of Spirit* and the notions of being-in-the-world, experience, and facticity. These convergences thus become strengths for a reconsideration of existentialism in the direction of a concrete dialectic open to intersubjectivity.

Building on this new theoretical line, Merleau-Ponty's subsequent works – *Humanism and Terror*, Adventures *of the Dialectic* – undertake the same effort: namely, to trace and phenomenologically bring to light the meaning of a lived dialectic that implies the overcoming of the ineliminable contradiction between the subjective and the objective, thereby becoming an ethical dialectic, a capacity to realize rationality. According to Merleau-Ponty, contemporary philosophers do not renounce rationality, the accord of oneself with oneself and with others, but only the pretense of a reason that is content with being right about itself and evades the judgment of others, proclaiming itself absolute reason.

Thus, seeking accord with ourselves and with others is precisely the opposite of irrationalism, which regards the contradictions of experience as insurmountable and our disagreement with others as inevitable. From this stems Merleau-Ponty's commitment to realizing a society where one is truly freer to the extent that others are free; and his insistence – directly stemming from Hegelian philosophy – on the necessity of a philosopher's presence that does not exhaust itself in mere conceptual activity or action aimed solely at itself. This presence is comparable to the figure of the servant, the "true hero" of the *Phenomenology of Spirit*, insofar as he

14 Therein, p. 134.

does not "believe in a genius of the Universe who prepares everything for his success and clearly indicates the way," but sacrifices his own life and challenges the established order in order to bring forth another, working and thus transforming "the world in such a way that in the end there is no more room for the master"[15].

Up to this point, we have examined the 1946 essay (*Hegel's Existentialism*) as the privileged site of the dialogue between Merleau-Ponty and Hegel. In reality, it must be said that this dialogue had already been undertaken by the French philosopher at the very beginnings of his reflection, in the 1938 work, where we already find several direct references to Hegel which later prove decisive for the ontological evolution of his Merleau-Pontian thought.

In *The Structure of Behavior*, Merleau-Ponty in fact proposes – so as to provide grounding for the philosophical implications present in the notion of behavior, and to explain the relation that binds the individual to the environment and the relation that is established between the physical, the vital, and the spiritual – precisely the concept of dialectic. This is evident in at least three fundamental passages:

1. In describing the physical or organic individual and its milieu, we have been led to accept the fact that their relations were not mechanical, but dialectical[16].
2. In recognizing that behavior has a meaning and depends upon the vital significance of situations, biological science is prohibited from conceiving of it as a thing in-itself *{en soi)* which would exist, *partes extra partes, in* the nervous system or *in* the body; rather it sees in behavior an embodied dialectic which radiates over a milieu immanent to it[17].
3. Aided by the notion of structure or form, we have arrived at the conclusion that both mechanism and finalism

15 Therein, p. 184.
16 Id., *The Structure of Behavior*, quot., p.160.
17 Therein, p.161.

> should be rejected and that the "physical," the "vital" and the "mental" do not represent three powers of being, but three dialectics.[18]

Here, dialectic *"guarantees"* the *"resolution"* of the oppositional tension, of the irreconcilable separation between the *"organic"* and the *"physical,"* between the *"vital"* and the *"psychical"* (in the sense of *"spiritual"*). But – one might object – by following this path (that is, from the perspective of an idea of overcoming), did Merleau-Ponty not run the risk of falling into the embrace of a totalizing dialectical movement which, precisely as such, would sacrifice the lower in favor of the higher?

> Physical nature in man – explains the author – is not subordinated to a vital principle, the organism does not conspire to actualize an idea, and the mental is not a motor principle *in* the body; but what we call nature is already consciousness of nature, what we call life is already consciousness of life and what we call mental is still an object vis-a-vis consciousness. Nevertheless, while establishing the ideality of the physical form, that of the organism, and that of the "mental," and *precisely because we did it,* we could not simply superimpose these three orders; not being a new substance, each of them had to be conceived as a retaking and a "new" structuration of the preceding one. From this comes the double aspect of the analysis which both liberated the higher from the lower and founded the former on the latter. It is this double relation which remains obscure (…).[19]

Here, already in his first work, the sense Merleau-Ponty intended to give to the notion of dialectic is made explicit. There is, however, something even more important to emphasize: the philosopher is already speaking here of an *"embodied dialectic,"* of a *"recurrence of the previous,"* of a *"double relation that remains obscure."* These are concepts

18 Therein, p.184.
19 Therein, p.184.

that will come to constitute the philosophical reflection of the later Merleau-Ponty – one need only think of the ontology of *la chair*, the figure of the *chiasm*, the notion of ambiguity.

At this point, it is useful to recall three *Working Notes* – dated respectively November 1, 1959; June 1960; and November 1960 – which seem to lead us directly back to the passage from *The Structure of Behavior* cited above:

> CHIASM – The chiasm (…) is also an exchange between me and the world, between the phenomenal body and the "objective" body, between the perceiving and the perceived: what begins as a thing ends as consciousness of the thing, what begins as a "state of consciousness" ends as a thing. One cannot account for this double "chiasm" by the cut of the For Itself and the cut of the In Itself[20].
>
> FLESH – MIND – *Define* the mind as the *other side* of the body – We have no idea of a mind that would not be *doubled* with a body, that would not be established on this *ground* – The "other side" means that the body, inasmuch as it has this other side, is not describable in *objective* terms, in terms of the in itself— that this other side is really the other side *of the body, overflows* into it (*Ueberschreiten*), encroaches upon it, is hidden in it— and at the same time needs it, terminates in it, is *anchored* in it. There is a body of the mind, and a mind of the body and a chiasm between them. The other side to be understood not, as in objective thought, in the sense of another projection of the same flat projection system, but in the sense of *Ueberstieg* of the body toward a depth, a dimensionality that is not that of extension, and a transcendence of the negative toward the sensible[21].
>
> VISIBLE – SEER – There is no coinciding of the seer with the visible. But each borrows from the other, takes from or encroaches upon the other, intersects with the other, is in chiasm with the other. In what sense are these multiple chiasms but one: not in the sense of synthesis, of the originally synthetic unity, but always in the sense of *Uebertragung,* encroachment, radiation of being, therefore. […] the double inscription outside and inside.[22]

20 Id., *The Visible and the Invisible,* quot., p.245.
21 Therein, p.259..
22 Therein, p.261

Even if Merleau-Ponty's language in this final phase of his speculative itinerary departs in form from that of the beginnings of his reflection, it nevertheless seems to us that the principal themes developed in *The Structure of Behavior* in fact lead to the questions taken up, repeatedly, in his later writings – confirming the coherence and organic continuity of the development of Merleau-Ponty's research. To support this thesis, it is possible to make one further reference, recalling the description of the essence of organic life – defined, in Hegelian terms, as the appearing of an interior in the exterior – which, while in *The Structure of Behavior* is presented as the inaugural act of a "fragment of extension" capable of manifesting an inner being outwardly, will, in *The Visible and the Invisible*, mark the ontological arrival at the idea of the *dehiscence of the flesh*[23]. Moreover, the very deconstruction carried out by Merleau-Ponty in the 1938 work with regard to the schools of experimental psychology should be understood not merely as the starting point for a subsequent theoretical development, but rather as the place where his philosophical position emerges with the greatest clarity. Thus, the tasks that Merleau-Ponty sets for himself from the very Introduction to *The Structure of Behavior* – "to understand the relations of consciousness and nature"[24], to explain "the vital meaning"[25] of the relations "of the organic individual and its milieu"[26], to give an account of that "phenomenon of life" appearing when "a piece of extension [...] turn(s) back upon itself and (begins) to express something, to manifest an interior being externally"[27]- can be matched and interpreted in light of the subsequent endo-ontology, or intra-ontology. In his Working Note dated January 20, 1960, Merleau-Ponty writes

23 See Mancini, S., *Sempre di nuovo. Merleau-Ponty e la dialettica dell'espressione*, Franco Angeli, Milan 1987, pp. 158-159.
24 Merleau-Ponty, M., *The Structure of Behavior*, quot., p. 3.
25 Therein, p.140.
26 Therein, p.148.
27 Therein, p.162.

that "Endo-ontology (...) is a going beyond the ontology of the In itself sé"[28] and therefore of the "view of the world from without, from the point of view of a Kosmotheoros"[29]. Moreover, he adds:

> I must no longer think myself *in the world* in the sense of the objective spatiality, which amounts to auto-positing myself and installing myself in the *Ego uninteressiert* -What replaces causal thought is the idea of transcendence, that is, of a world seen within inherence in this world, by virtue of it, of an Intra ontology, of a Being encompassing-encompassed, of a vertical, dimensional Being, dimensionality -. And what replaces the antagonistic and solidary reflective movement (the immanence of the "idealists") is the fold or hollow of Being having by principle an *outside,* the architectonics of the configurations.
>
> There is no - consciousness
- projections
- In itself or object
>
> There are fields in intersection, in a field of fields wherein the "subjectivities" are integrated, as Husserl indicates in the unpublished text on teleology and the phenomenological absolute, since they bear in their intrastructure a *leistende subjectivität* which is wholly *supported* on them.[30]

The question addressed by Merleau-Ponty in *The Structure of Behavior* resurfaces, like an underground current, in his posthumous work and becomes its very cornerstone, the privileged site from which the philosopher believes it necessary to continue interrogating. The dialectic of the inside and the outside (du dedans et du dehors) represents the entire arc of Merleau-Ponty's reflection and traverses all of its phases: the analyses of oppositional figures (exteriority/interiority, body/soul, human/world, figure/ground, sense/non-sense, visible/invisible, philosophy/non-philosophy) all move in

28 Id., *The Visible and the Invisible,* quot., p.226.
29 Therein, p.227.
30 Ibid.

this single direction: the search for a *Being of indivision*[31], of *promiscuity*[32], of *overlapping*[33] which is always in a relation of junction, adherence, and belonging such that it allows neither overlooking nor fusion, but – at once – distance and proximity. This is why, even for the Merleau-Ponty of *The Structure of Behavior*, what must be understood is not so much the subject-object relation, but rather that *dehiscence* or transcendence which is Being itself – that is, the relation with Being that takes place within Being itself.

At this point, we can fully grasp the meaning of the *Working Note* dated November 1959, in which Merleau-Ponty writes:

> Consider the right, the left: these are not simply contents within a relational spatiality (i.e. *positive)*: they are not *parts* of space (Kant's reasoning is valid here: the whole is primary), they are total parts, cuts in an encompassing, topological space – Consider the *two,* the *pair,* this is not *two acts, two syntheses,* it is a fragmentation of being, it is a possibility for separation (two eyes, two ears: the possibility for *discrimination,* for the use of the diacritical), it is the advent of difference (on the ground of *resemblance* therefore, on the ground of the omou hn panta).[34]

The *"total part"* of *"fragmentation," "discrimination,"* and *"difference"* undermines from within the two-thousand-year-old Western philosophical tradition that had created the myth of the *Kosmotheorós* and establishes itself as the "*the heart of being*"[35]. But this is certainly not a recovered immediacy of a real fusion; rather, it is only a partial coincidence, a *"coincidence from afar,"* and therefore a constitutive gap that incessantly differentiates into *"from-within"* and *"from-without."* A reversibility that is, by essence, unfinished,

31 Therein, p. 255.
32 Therein, p. 270.
33 Therein, p. 246.
34 Therein, p. 216-217.
35 Therein, p. 221.

"*always imminent and never realized in fact*"[36], because "it's realization would be, by that very fact, the loss of the world, the closure within an inside without an outside" and because "there is no experience that is not a metamorphosis of an inside into an outside and of an outside into an inside." Therefore, Being can only be "multidimensional and polymorphic" and nothing positive.

In light of these considerations, which certainly represent a decisive point for understanding the late Merleau-Ponty, one can also read a Working Note from May 1960, entitled:

> Metaphysics – Infinity
> World – Offenheit
> World and Being:
>
> their relation is that of the visible with the invisible (latency) the invisible is not another visible ("possible" in the logical sense) a positive only *absent* It is *Verborgenheit* by principle i.e. invisible *of the visible, Offenheit* of the *Umwelt* and not *Unendlichkeit* – *Unendlichkeit* is at bottom the *in itself,* the *object* – For me the infinity of Being that one can speak of is *operative,* militant finitude: the openness of the *Umwelt* – 1 am against finitude in the empirical sense, a factual existence that *has limits,* and this is why I am for metaphysics. But it lies no more in infinity than in the factual finitude.[37]

The recognition of the inability of the ontological dualism between *in-itself* and *for-itself* to account for the most common phenomena of human existence thus led Merleau-Ponty onto another path: that of a thought of the *entre-deux*, of ambiguity, of laterality. By rejecting a concept of dialectic that would impose violence on phenomena, reducing them all to a form of "*pure*" opposition of contraries, and instead embracing the weight of a thought of difference internal to dialectic

36 Therein, p. 147.
37 Therein, p. 251.

itself, the French philosopher offers us a novel analysis and a new meaning for the *"traditional"* concepts of existence, negativity, and questioning.

Later, we will see how this commitment to understanding unity from difference, and the insistence on the essentially unfinished character of every synthesis, leads Merleau-Ponty to deepen and revise even the ideas of history, freedom, and the sedimentation of facts. Existence, far from being considered merely as the mode of being of man, is an ontological category; it is the paradoxical identity (in Heideggerian terms) of *Existenz* and *Instandigkeit*, of *ek-stase* and *in-stance*, of stepping out of oneself and remaining close.

It becomes clear that the true thought of the exteriority of being is not a thought of separation (in the sense of the dichotomy *In-itself – For-itself*), but, on the contrary, a thought of *être-dans* or *en-être*, of existential immanence – that is, of the relation or, in Merleau-Pontyan terms, the *carnal junction* between self and world. Here, the negative already changes in sense and significance. This marks, in parallel, Merleau-Ponty's increasingly pronounced distancing from Sartre. It is against these positions, in fact, that a *Working Note* from just a few months earlier is directed:

> The problem of negativity is the problem of depth. Sartre speaks of a world that is not vertical, but in itself, that is, flat, and for a nothingness that is absolute abyss. In the end, for him depth does not exist, because it is bottomless – For me, the negative means absolutely nothing, and the positive neither (they are synonymous) and that not by appeal to a vague "compound" of being and nothingness, the structure is not a "compound." I take my starting point where Sartre ends, in the Being taken up by the for Itself- – It is for him the finishing point because he starts with being and negentity and *constructs* their union. For me it is structure or transcendence that explains and being and nothingness (in Sartre's sense) are its two abstract properties. For an ontology from within, transcendence does not have to be constructed, from the first it is, as Being doubled with nothingness, and what is to be explained is its doubling (which, moreover, is

> never finished) [...] For Sartre, it is always I who forms depth, who hollows it out, who does everything, and who closes from within my prison in upon myself – For me, on the contrary, (...) this is not a non-being that makes itself be.[38]

As we can see, Merleau-Ponty's opposition to a *"positive"* concept of negativity is decisive – that is, to a negative that is negative only by mirror opposition to the positive, and which therefore confirms itself simply as a new positive with its sign reversed. This opposition is decisive precisely because the philosopher wishes to maintain his discourse on an ineliminable *"ambiguity,"* which, while preventing the attainment of a final synthesis or reconciliation, likewise prevents any schism or opposition that would erase the meaning of the relation.

Thus, in every Merleau-Pontian work, we witness the confrontation of two opposing terms, irreducible to one another – figure/ground, body/world, sense/non-sense, eye/spirit, visible/invisible – which are not resolved into a *tertium*, but neither do they remain unrelated, precisely because each term lives from the relation that simultaneously brings it closer to and distances it from the other. Here is the dual movement, the *chiasm*, the *"relational noose,"*[39] dialectic as an *"ascension that takes place in situ."* In fact, there is no figure without a ground, no visible without the invisible, no body that is not of and in the world. Yet all of this has given rise to a decisive misunderstanding: figure, the visible, and being have been defined as positives emerging from an absence, from a negative, from nothing. That is, the positive that springs from the negative, that acquires being from non-being. Hence the hasty definition of an ontology which, far from being understood as Merleau-Ponty intended – that is, negative and indirect – is, in fact, returned to the canons

38 Therein, p. 237.
39 Sini, C., *Il silenzio e la parola*, Marietti, Genova 1989, p. 31 (my translation)

of traditional metaphysics: the ontology of *la chair* and the ontology of the plenum, of pure positivity, of fullness.

This interpretation is even more incomprehensible if one recalls the statement contained in the *Working Note* of February 1960, cited earlier, there is *"not a non-being that makes itself be."* It therefore seems appropriate, and indeed necessary in order to avoid any possible interpretive misunderstanding regarding the ontology of *la chair*, to return to the question of the negative, starting precisely from Merleau-Ponty's rigorous critique of Sartre's *Being and Nothingness*. In the second chapter of *The Visible and the Invisible*, entitled *Interrogation and Dialectic*, the French philosopher undertakes a careful examination of Sartre's work, highlighting all the limits and internal contradictions of an ontology of the pure Subject.

> The thought of the pure negative or of the pure positive is – according to Merleau-Ponty – [...] a high-altitude thought, which operates on the essence or on the pure negation of the essence, on terms whose signification has been fixed and which it holds in its possession. Sartre does indeed say that *at the end of his book* it will be permissible to move to a broader sense of Being, which contains Being and nothingness. But this is not because the initial opposition would have been overcome; it remains in all its rigor, it is that initial opposition that justifies its own reversal, that triumphs in this defeat; the passion of the For Itself, which sacrifices itself in order that being be, is still its own negation by itself. It is tacitly understood that from one end of the book to the other we are speaking of the same nothingness and of the same being, that one unique spectator is witness to the progress, that he is not himself caught up in the movement, and that inasmuch as that is so the movement is illusory. A negativist or positivist thought rediscovers that postulate of the philosophy of reflection that no result of the reflection can retroactively compromise him who operates the reflection nor change the idea we form of him for ourselves. And it cannot be otherwise if one starts with the pure negative: for it will never admit anything into itself, and even if one comes to recognize that it has need of Being, it will need Being only as a distant environment that does not adulterate it. It will dispose it about itself, as a pure spectacle or as what it

> has to be, it will elevate it to truth or to signification; but it will itself remain the nothingness it was, its devotion to Being will confirm it as nothingness. The negativist (or positivist) thought establishes between nothingness and being a massive cohesion, both rigid and fragile at the same time: rigid since they are finally indiscernible, fragile since they remain unto the end absolute opposites.[40]

Merleau-Ponty's effort is thus to overcome the demands of reflective philosophy still present in Sartre. Guided by this effort, almost as if speaking to himself, the philosopher writes, in a marginal note – also from the posthumous work: "I said in tum that "nothingness is not" and "being is" are the same thought – and that nothingness and being are not united. Connect the two: they are not united precisely because they are the same thing in two contradictories = ambivalence."[41] Thus returns, reworked and deepened, the initial concept of *"ambiguity,"* now intensified as ambivalence, constitutive doubleness, the *chiasm* of Being and nothingness – all terms aimed at escaping the vicious circle of an absolute negativism which, if taken to its extreme consequences, turns into absolute positivism with the risk – as Merleau-Ponty emphasizes, critiquing Sartre – to restore the philosophy of identity, to ignore "density, depth, the plurality of planes, the background worlds"[42]. Since it is precisely the claim "to name this nothingness which he is, to coincide with this fissure in being "that constitutes "a variant of the principle of identity which defines being"[43].

> In thinking on the basis of the pure negative we already decide to think according to identity; we are already in identity, since this negative which nothing can limit in its own order, having to go on to the limit of itself, will be also, and fundamentally, a

40 Merleau-Ponty, M., *The Visible and the Invisible,* quot., pp. 69-70.
41 Therein, p. 69.
42 Therein, p. 68.
43 Therein, p. 67.

> negation of itself, and therefore will be pronounced in the form of an advent of pure being. There is a trap inherent in the thought of the negative: if we say that it is, we destroy its negativity; but if we maintain strictly that it is not, we still elevate it to a sort of positivity, we confer upon it a sort of being, since through and through and absolutely it is *nothing* [...] A negativist thought is identical to a positivist thought.[44]

In summary, the formula for negativistic thought compensates "for one abstraction with a counter-abstraction"[45] remaining, in fact, on the level of an a-dialectical thought that leaves behind the terrain of pre-predicative experience, of the spontaneous engagement with the lived world, of the irreducible distance between the unre-flected and reflection.

This is how, from the critique of Sartrean negativism, Merleau-Ponty draws new impetus for the delineation of a thought that, while remaining faithful to the negative, always operates in the mode of interrogating mute experience, recognizing at the same time the original nexus of the primordial insertion of the self into the world and their reciprocal, unbridgeable difference. The *Weltthesis* is, in this light, the initial acknowledgment that there is something (*Etwas*), the ontological prominence of *la chair*, the invisible framework that gives the visible to the visible.

In this way, Merleau-Ponty overcomes the sterile oscillations of Sartrean philosophy and proposes a genuine dialectical conception of Being, in which being and nothingness are realized as part of the same movement: the negative is reintegrated into Being itself as a simple constitutive element. From here opens the path toward a dialectical conception of *chair*. Here, *the chair* is not synthesis, coincidence, return to the identical, positivity, or substance, but rather reversibility, reciprocal action, non-coincidence, latency, dimensionality,

44 Therein, pp. 67-68.
45 Therein, p. 68.

dehiscence, the fabric of difference, a u-topical element, being at a distance, contingency.

Thus, following the resolute critique of reflective philosophy and Sartre's negativistic philosophy, Merleau-Ponty proposes a third way: a philosophy that cannot consist in a "rupture with the world, nor in coincidence with it, nor in an alternation of rupture and coincidence, but simply in openness"[46]. The openness to the natural and historical world is neither an illusion nor an *a priori*; it is our very involvement in Being. The disclosure of Being cannot occur except through "a relation to Being through a being"[47] beyond the distinction between Being-for-itself and Being-in-itself, between the intuition of Being and the *"non-intuition"* of nothingness. This allows Merleau-Ponty to radicalize the Husserlian concept of *Lebenswelt* and to arrive at the notion of *la chair* as *Être brut*, *Être sauvage*. This raw, polymorphic being is not a constituted something, it is not a new *in-itself*, but rather "the flesh of things that speaks to us of our own flesh"[48]. "Here is the common tissue of which we are made [...] And the perception of this perception (the phenomenological "reflection") is the inventory of this originating departure whose documents we carry in ourselves, of this *Ineinander* that awakens to itself, it is the usage of the *immer wieder* which is the sensible, the carnal itself"[49]. It is precisely thanks to the carnal element that there can be not only the self and the object, but also their reciprocal, inexhaustible seeing; a seeing that is ontologically prior because it is a "vision of the origins", "pre-intentional" and which "articulates itself within us, unbeknown to us"[50]. As Merleau-Ponty writes again in a working note from November 1959, "the visible (...) is the *Urpraesentation* of

46 Philippe, M.-D., *Une philosophie de l'Être est-elle encore possible?* Tequi, Paris 1975, p. 111 (my translation)
47 Merleau-Ponty, M., *The Visible and the Invisible,* quot., p.118.
48 Therein, p. 193.
49 Therein, pp. 203-204.
50 Therein, p. 208.

the *Nichturpraesentierbar* – *To see* is precisely, in spite of the infinite analysis always possible, and although no *Etwas* ever remains *in our hands,* to have an *Etwas*". And he adds, "Is this then a pure and simple contradiction? Not at all: the visible ceases to be an inaccessible if I conceive it, not according to the proximal thought, but as an encompassing, lateral investment, *flesh.*"[51]. What then deserves the name of Being is not, strictly speaking, the horizon of *pure* being, but the system of perspectives that introduces us to Being itself—that is, the movement by which being is never fully unfolded before me, but opens itself at the intersection of my viewpoints and those of others. For this reason, the sensible world and the historical world are always *"inter-worlds,"* places, so to speak, anonymous because constantly confused, intertwined within the generality and promiscuity of *la chair* according to the modes of difference and divergence, yet implicated (hence the concept of Being as encompassing) in the cohesion of a single *something* (*Etwas*).

Only at the end of this path does Merleau-Ponty affirm that the *Être brut* or *sauvage* is the perceived world, the "*être déjà là",*[52] the pre-critical[53]. This is not to be understood in the sense of negation or in relation to our knowledge, but as the radical openness that is both distance and transcendence. Beyond the analysis of the intentionality of knowledge (that is, the critical aspect), there exists a fundamental, massive, ontological unity: the *il y a*, the inaugural *there is*, the common fabric of which we are made, the *être sauvage*, the *chair du monde*. Here, the flesh of the world and Being-seen signify the same thing: a Being that is eminently perceivable and through which alone one can understand perceiving, the act of perceiving, since my

51 Therein, p. 217.

52 Therein, p. 202.

53 See Ales Bello A., *"Essere grezzo" e hyletica fenomenologica. L'eredità filosofica del visibile e l'invisibile,* in "Chiasmi International", n. 10, 2008, pp. 139-161.

body is a perceiver because Being exists; "not Being in itself, identical to itself, in the night, but the Being that also contains its negation, its *percipi*"[54].

With this final outcome, so to speak, the circle of Merleau-Ponty's philosophy closes, reconnecting with its beginnings and showing what was already implicit in *The Structure of Behavior*: the ontological character of perception. Indeed, as the philosopher explains:

> Perception opens the world to me as the surgeon opens a body, catching sight, through the window he has contrived, of the organs in full functioning, taken *in their activity,* seen sideways. It is thus that the sensible initiates me to the world, as language to the other: by encroachment, *Ueberschreiten.* Perception is not first a perception of *things,* but a perception of *elements* (water, air...) of *rays of the world,* of things which are dimensions, which are worlds, I slip on these "elements" and here I am in the *world,* I slip from the "subjective" to Being.[55]

But what, more precisely, does this passage from the subjective to Being consist of? It consists in the very structure of the *chiasm*, a dynamic structure of interweaving, reversibility, and mirroring, in which the concept of negativity is integrated, since the internal dynamism of the chiasm indeed generates within itself a depth, a void, an invisible that are not nothing, but rather the essential elements of the life of *la chair*. This void and depth represent, strictly speaking, precisely what Hegel calls the *"work of the negative"* – not a mere conceptual abstraction, but an operative negativity inherent to the very life of *la chair*.

The philosopher's proposal is therefore clear: an open, endless dialectical process, without synthesis, without a totalizing and all-encompassing system. This process, which Merleau-Ponty will call *"hyper-dialectic,"* definitively settles

54 Merleau-Ponty, M., *The Visible and the Invisible,* quot., pp. 250-251.

55 Therein, p. 218.

accounts with the myth of the spectator consciousness and is configured simply and essentially as *Offenheit*.

Yet, the *"openness"* of which our thinker speaks goes decisively beyond the Hegelian interpretation of dialectic. The space in which it unfolds is here fragmentation, divergence, the advent of difference. It is precisely the dialectical movement of *Offenheit* as the transcendence of identity in difference that marks Merleau-Ponty's distance from Hegel – though it is with that Hegel that the French philosopher will continue to dialogue until the very end.

§ 2. *The Offenheit of the Umwelt*

During 1952–53, in the lecture entitled "The Sensible World and the World of Expression," Merleau-Ponty, in conclusion, set out "to establish definitively the philosophical meaning (…) of the relation between nature and culture". Only in this way, in fact, it will be possible to decide "whether the dialectic of expression implies that a mind is already present in nature, or that nature is immanent to our mind, or [else] to search instead for a third philosophy beyond this dilemma"[56]. The need to "search for a third philosophy", which reappears in the pages of *The Visible and the Invisible*, precisely where the philosopher speaks of *"indirect ontology"* and *"negative philosophy,"*[57] is not only supported by the aim of replacing the "history of philosophy" tout court with a "history of non-philosophy" as "pure interrogation,"[58] but becomes –

56 Merleau-Ponty, M., *The Sensible World and the World of Expression: Course Notes from the Collège de France, 1953*, transl. by B. A. Smyth, Northwestern University Press, Evanston, Illinois 2020, pp. XII–XIII and XXXVI.

57 Merleau-Ponty, Maurice, *The Visible and the Invisible*, quot., p. 179.

58 Id., *Themes from the Lectures at the Collège de France, 1952-1960*, quot., p.102.

in Merleau-Ponty's final university course at the Collège de France in 1961, entitled *"Philosophy and Non-Philosophy since Hegel"* – he starting point for a new and more mature dialogue with Hegel. In this perspective, the necessity of opening oneself to that "possibility of philosophy"[59] emerges, which is the very essence of thinking, thus recovering the sense "of what Husserl called a 'poetry of the history of philosophy'"[60].

The 1961 course thus acquires its full significance because it highlights – certainly more clearly than the critique of reflective philosophy outlined in *The Visible and the Invisible* – the project of finding, in the history of philosophy, at the moment when it becomes, with Hegel, a philosophy of history, a passage beyond its own limits. This project simultaneously reveals the phenomenological traits of both the Hegelian and Husserlian trajectories. Indeed, the intention to critically mediate the Hegelian dialectic of the Absolute with Husserl's transition "from 'philosophy as a strict science' to philosophy as pure interrogation"[61] is emphasized. In this context, the key notions introduced in several essays of *Signs*, in *The Prose of the World*, *Eye and Mind*, and *The Visible and the Invisible* – such as *flesh*, *reversibility*, *dehiscence*, *circularity*, *articulation*, and so forth – enter into a tacit agreement with reflective philosophy, which is now compelled to give way to a different dimension of philosophizing: it is no longer a question of establishing or

> understand how a for-itself can think of another from the ground of its own absolute solitude or how it could think of a preconstituted world in the very moment that it constitutes the world: the inherence of the self-in-the-world or of the world-in-the-self, what Husserl calls the Ineinander, is silently inscribed in an all-embracing experience which composes these incompossibles,

59 Therein, p. 99.
60 Therein, p. 114.
61 Therein, p. 105.

> and philosophy becomes the enterprise of describing, outside of the logic and vocabulary at hand, the universe of living paradoxes. The reduction no longer involves a, return to ideal being, but brings us back to the spirit of Heraclitus, to an interweaving of horizons, to an open Being. Through having "forgotten" the flux of the natural and historical world, having reduced it to the constructions of the objectivity of the natural sciences, reason and philosophy have become incapable of understanding and so of mastering man's historical fate. The horizon of "infinite tasks" opened up in the sixteenth and seventeenth centuries has dropped out of sight, while reason and philosophy have compromised with an ideal of objectivism which made knowledge of the mind and history an impossibility.[62]

The *"forgetting"* that Merleau-Ponty reproaches in traditional philosophy can only be overcome by a new reflection capable of opening itself to that pre-predicative experience *"removed"* by traditional thetic reason. Here, Heraclitus' yuch′, to which Merleau-Ponty refers in the cited passage, is certainly the explicit manifestation of that *pure interrogation* that the late Husserl indicated as the project of a new philosophy: "You will never find the boundaries of the soul, even if you follow every road; so deep is its ground". And as Husserl continued: "(…) every 'ground' [Grund] that is reached points to further grounds, every horizon opened up awakens new horizons, and yet the endless whole, in its infinity of flowing movement, is oriented toward the unity of one meaning; not, of course, in such a way that we could ever simply grasp and understand the whole"[63]. Thus, our relationship to the world is inexhaustible and is characterized as surpassing and loss, as tension *toward*, as non-coincidence. But philosophy's renunciation of possession of the truth does not in the least mean surrender, weakening of thought, or

62 Therein, pp.108-109.

63 Husserl, E., *The Crisis of European Sciences and Transcendental Phenomenology: An Introduction to Phenomenological Philosophy*, quot.

betrayal of its task. It simply signifies the constant pursuit of another truth of philosophy, which perhaps also wishes to be non-truth, non-foundation. This pursuit therefore seeks "to recover through an absolutely new way of thinking the experience which underlies metaphysics."

> It can be said of this Being – the rose, said Angelus Silesius, is "without any why," blooms because it blooms, the rose-spectacle, the rose-totality – that is has no cause outside of itself and moreover is not the cause of itself, that it is without ground, being the absence in principle of any ground. What more can be said of this radiation of qualified being, of this active being, or action of "standing-in" [....] Unlike other words, the word to be is not a sign to which one could find a corresponding "representation" or object: its meaning is not distinct from its operation, which is to make Being speak in us rather than us speak in Being. For how would we speak of Being, since those beings and shapes of Being, which open to us the only conceivable access to it, at the same time hide it form us by their mass, and since every unveiling is simultaneously a dissimulation? What has been called the "mystique" of Being [....] is the effort to integrate truth with our capacity for error, to relate the incontestable presence of the world to its inexhaustible richness and consequent absence which it recuperates, to consider the evidence of Being in the light of an interrogation which is the only mode of expressing this eternal elusion.[64]

The radicality of this task-pursuit can already be traced back to the beginnings of Merleau-Ponty's reflection. Indeed, in the 1945 essay *"Metaphysics and the Novel"*, the problem of the truth of philosophy, even if still framed at the level of the experience of consciousness, was already configured as the central question of thinking. From that moment, Merleau-Ponty was fully convinced that language bore witness to an implication in the world that it itself sought to describe and that, moreover, continuously eluded it, being itself an integral

64 Merleau-Ponty, M., *Themes from the Lectures at the Collège de France, 1952-1960,* quot., pp. 110-111.

part of that implication, of that paradoxical *en-être*. The impossibility of a direct thematization of that entering into oneself which is also an exiting from oneself, and vice versa – the continual failure and the illusory nature of an overview knowledge capable of accessing a total and residue-free truth – brings the *novel* close to *metaphysics*, non-philosophy close to philosophy.

> Everything changes – writes Merleau-Ponty – when a phenomenological or existential philosophy assigns itself the task, not of explaining the world or of discovering its "conditions of possibility," but rather of formulating an experience of the world, a contact with the world which precedes all thought about the world. After this, whatever is metaphysical in man cannot be credited to something outside his empirical being-to God, to Consciousness. Man is metaphysical in his very being, in his loves, in his hates, in his individual and collective history. And metaphysics is no longer the occupation of a few hours per month, as Descartes said; it is present, as Pascal thought, in the heart's slightest movement. From now on the tasks of literature and philosophy can no longer be separated. When one is concerned with giving voice to the experience of the world and showing how consciousness escapes into the world, one can no longer credit oneself with attaining a perfect transparence of expression. Philosophical expression assumes the same ambiguities as literary expression, if the world -is such that it cannot be expressed except in "stories" and, as it were, pointed at.[65]

Philosophy must become non-philosophy, speaking also of the "shadows crowded with creatures", of the "shows performed without an audience", of the sense of an "old jacket lying on a chair (…) in the silence of a country house," that "idoes not know what it is;", while "I am always secretly aware that there are other things in the world beside me and what I see"[66]. It is therefore necessary for the philosopher to abdicate his role as *Kosmotheoròs* and to engage with another

65 Id., *Sense and Non-sense*, quot., pp. 27-28
66 Therein, pp.28-29.

mode of speaking, one that allows us to understand that if "every object can affirm its existence only by depriving me of mine"[67], this is because my involvement with Being (the Ineinander) is always at the limits of perfect tangibility.

"*Philosophie ou non-philosophie*?"[68]. Lefort's question echoes the clarification expressed by Merleau-Ponty at the beginning of the course: "No battles occur between philosophy and its adversaries. Rather what happens is that philosophy seeks to be philosophy while remaining non-philosophy, i.e. a 'negative philosophy' (in the sense of 'negative theology'). 'Negative philosophy' has access to the absolute, not as 'beyond,' as a positive second order, but as another order which must be on this side, the double – inaccessible without being passed through. True philosophy scoffs at philosophy, since it is a-philosophical."[69] The idea pursued by Merleau-Ponty – already implicitly expressed by Hegel, Marx, Kierkegaard, Nietzsche, Heidegger – is therefore that of a philosophy of non-philosophy, one that finally establishes and maintains a close connection with what classical philosophy had always devalued: the *beyond-here*, existence, life, experience, and above all the concept of the Absolute as the *"non-separated"*, as *"being-present-to"*.

The philosophy of non-philosophy, characterized as a" philosophy of several entries"[70], thus rediscovers the key to the Absolute precisely in the *Phenomenology of Spirit*. The French philosopher takes up the principle enunciated by

67 Therein, p.29.

68 Lefort, C., *Philosophie et non-Philosophie*, in "Esprit", n. 6, 1982, p. 101. See also, in particular: Dastur F., *Merleau-Ponty et Hegel. Ontologie et dialectique,* in "Chiasmi International", n. 11, 2009, pp. 33-47; Hirose K., *Instituer le chiasme: à partir du cours sur Hegel de Maurice Merleau-Ponty,* in "Chiasmi International", n. 16, 2014, pp. 221-237; Vuillerod J.B., *Merleau-Ponty hégélien?,* in "Chiasmi International", n. 19, 2018, pp. 101-114.

69 Silverman, Hugh J., *Philosophy and Non-Philosophy since Merleau-Ponty*. Routledge, London New York 1988, p. 9.

70 Merleau-Ponty, M., *The Visible and the Invisible*, quot., p. 174.

Hegel: "one attains the absolute by way of a phenomenology (the appearance of mind; mind in the phenomenon). This is not because the phenomenal mind is on one level of a scale, after which one moves on to the absolute, but because the absolute would not be absolute if it did not appear as absolute."[71] Therefore, in this sense, true philosophy does not "end up fulfilling the need to write ideas across the sky in cosmic capital letters", but is grounded onto experience. "In being a full phenomenon, the absolute is fully absolute."[72] For the later Merleau-Ponty, philosophical inquiry is possible only if it is carried out in the sense of a *dialectical phenomenology*, insofar as it does not consist in an access to the Absolute by means of knowledge, but rather reveals the presence of the Absolute as it manifests itself in the thing, in the phenomenon. Hegel's indication of the originary relation between consciousness and Absolute, between spirit and nature, explains – according to the French philosopher – the affirmation which, at first sight, could seem to be a "dogmatic leap," namely: *"the Absolute alone is true, or the True alone is absolute."*

The non-demonstrability of this proposition makes the originary character of the relation emerge: the Absolute does not *prove* itself, but *self-shows* in its phenomenal becoming as a "*membrure*" of experience, as a pure *parousia*. There is, therefore, writes Merleau-Ponty:

> This implies that the absolute within us is internal to the truth, that there is an absolute which is nothing other than our truth, and which is not separated from 'understanding.' The attitude here is not dogmatic. On the contrary, it is the true radical, critical attitude. By it, we commit ourselves to 'giving' the concept of the absolute, 'understanding,' the objective, the subjective 'to give this concept' (Holz, p. 108; trans., p. 11) rather than presupposing what everyone knows or assuming the concept that we already have of it. We begin with the true as absolute and with

71 Silverman, Hugh J. *Philosophy and Non-Philosophy since Merleau-Ponty*. quot., p. 9.

72 Therein, p. 305.

> the absolute as true. We start from a knowledge [*connaissance*] which can only form a unity with our being as a knowing being [*étre sachant*]. This knowledge cannot correspond to 'science making its appearance' (Holz, p. 108; trans., p. 12) as a nexus of knowledge and absolute, of our Being and the Being. – We must grasp this nexus fully and not in the emptiness of 'representations' [*Vorstellungen*]".[73]

"*From the principle*: the absolute, an *und fur sich schon bei uns ware*", underlines Merleau-Ponty. "For if the Absolute were not already *with* us from the very beginning, how could we ever grasp it? Here the decisive importance of the *beginning* becomes evident – of the origin of truth, of its very *birth.* And this birth is not the passage toward another *Seiende*, given the identity between the Absolute and the True"[74].

> From the absolute to us, the whole intermediary is 'the ray of light through which the truth touches us' – the intermediary is not the 'difference between ourselves and this understanding.' [...] This does not signify a dogmatism (an external link: 'knowledge'-absolute) but on the contrary an internal and reversible relation (without hierarchy) between true knowledge and absolute knowledge. The absolute is in the nature of the knowledge, and knowledge is in the nature of the absolute. – What is given is their relationship, from which 'subject' – 'object' must be clarified.[75]

The task of the philosopher is therefore to begin from "the true as absolute and the absolute as true," seeking to grasp, in the very emergence of knowledge, the *chiasm* between

73 Therein, pp.16-17. Reference is also made to Heidegger, *Holzwege*, Frankfurt-am-Main, Klostermann 1950.

74 Carbone, M. *Alla ricerca dell'*a-filosofia. Merleau-Ponty e la *Einleitung*, alla *Phänomenologie des Geistes*, in VV.AA., *Negli specchi dell'Essere. Saggi sulla filosofia di Merleau-Ponty*, quot., p. 215 (my translation).

75 Merleau-Ponty, M., *Philosophy and Non-Philosophy since Hegel*, in Silverman, H. J., *Philosophy and Non-Philosophy since Merleau-Ponty*, Routledge, New York and London 1988, p. 30.

knowing and the absolute, between our being and Being. Here the Merleau-Pontyan demand to unify knowing and the absolute, the real and the rational, becomes evident, to ground theoretically the reciprocal and inseparable relation between ourselves and the absolute. Hence: a *vertical* conception of the world, the necessity of an *Erkennen* that is not distinct from our own being, the binding and interweaving of the *Gestalten*. "the identity of our being and knowing "[76] leads back to the primordial *Weltthesis* of existence and thus brings Hegel and Husserl together in a single direction: Merleau-Ponty seeks to re-situate the Hegelian concept of the Absolute at the center of the originary, pre-reflective life, at *Être brut*, at the *wild Logos*. Between the Absolute and mute, pre-categorial experience there is no distance; they are one and the same. Through this new concept of the Absolute, the French philosopher overcomes the paradox of reflective philosophy, placing himself outside its fundamental coordinates. He identifies the transcendental no longer in the *Cogito*, as the constitutive horizon of meaning, but in *life* – and more precisely in its principle of productivity, which will appear as the raw, antepredicative Being unfolding in the "flesh" of the world as *wild Logos*, that is, in excess of the legislative function of the *Cogito*. Hence the co-originarity, the circularity, between the pre-reflective and reflection, dialectic as "Self-manifestation, disclosure, in the process of forming itself"[77]. From this conception there emerges the intimate bond, the cohesion, the reversibility between *appearance* and *Absolute* that Merleau-Ponty discerns in the *Einleitung* to the *Phenomenology of Spirit*. This marks a "reversal" in the history of philosophy, since, if the Absolute is "the other side" of *Erscheinung*, this means that it must be sought not *behind* the phenomenon, but *within* appearing itself, in the phenomenon.[78] "Presence

---

76 Therein, p.15.
77 Merleau-Ponty, M., *The Visible and the Invisible*, quot., p. 91.
78 This point is also central in the article by Lisciani-Petrini, E., *La*

of the absolute and revealing the phenomenon are synonyms because the absolute is the subject, self-conscious-ness, and the phenomenon represents the becoming of this self-consciousness. The phenomenon and the absolute are linked, because every phenomenon represents the relation to itself, and because the absolute is itself, that is, relation to itself."[79] Therefore, "True philosophy does not go 'behind': behind the 'appearance' and behind the one which is in the world. Philosophy cannot conceive the absolute except as the other side of the 'appear-ance' or the phenomenon."[80]

It is in the wake of these considerations that Merleau-Ponty can interpret the famous Hegelian passage: "In my view, which can be justified only by the exposition of the system itself, everything turns on grasping and expressing the True, not only as *Substance,* but equally as *Subject*"[81] – as one of the earliest moments in the history of a-philosophy. The philosopher writes:

> Phenomenological theme (*erscheinende Wissen*, the blossoming of knowledge) seems to imply an overthrowing of philosophy. There is no absolute *Ansich*, and no absolute *fiir uns*, for the same reasons, i.e. their reciprocal relativization, their 'mutual intertwining' [Ineinander]. This means that the very dimension of the absolute is to conceive in a fashion other than the way philosophy has done until now. The absolute is neither 'free' knowledge (that is, without attachments, without an outside, without adherence to the in-itself), nor 'free' object (that is, purely an object, without adherence to knowledge), but the Ineinander of the two. The (ultimate) formula: 'the absolute is the subject' does

passione impolitica della politica. Merleau-Ponty tra Filosofia e non-filosofia, in VV.AA., *Nichilismo e politica*, Laterza, Rome-Bari 2000, pp. 55–73.

79 Merleau-Ponty, M., *Philosophy and Non-Philosophy since Hegel.* quot., p.18.

80 Therein, p.39.

81 Hegel, G W F. *Phenomenology of Spirit,* quot., pp. 9-10.

> not mean that there is only a subject. […] The absolute is the 'sameness' of the absolute and the relative.[82]

Yet precisely because the Absolute is a "phenomenon," it escapes the duality of subject and object and thereby discloses the ontological features of phenomenology – features that, in his 1945 work, had not yet been fully thematized. The pivotal moment in this ontological development – which simultaneously shapes the epistemological dimension of his reflection is captured in a specific note by Merleau-Ponty, where he raises the question of the relation between phenomenology and absolute knowledge: that is, between what appears and the act of apprehending or taking possession of what appears.

> Phenomenology is this self-presentation of mind, an appearance which is not an effect of the absolute, but the absolute itself. In that respect, philosophy is experience. […] We must understand the relation between 'understanding' and the absolute as given in our life (hence an absolute which will also be 'understanding'). We must truly recast the concepts of subjectivity and objectivity, absolute and knowledge according to their contact with our life. We must take science within us in its nascent state; 'science, in that it appears, is an appearance itself' […] *Erkennen* is not to be placed outside the absolute because the absolute is not to be placed outside. […] The distinction between the understanding [*connaissance*] and the absolute (the critical attitude) is a false radicalism.[83]

Returning to the title of the first part of the Phenomenology of Spirit – *Science of the Experience of Consciousness* – the French philosopher demonstrates his adherence to the idea that the experience of consciousness constitutes the praxical itinerary in which the Absolute comes to presence: the nexus

82 Merleau-Ponty, M., *Philosophy and Non-Philosophy since Hegel*. quot., p. 39.
83 Therein, pp. 15-16.

between consciousness and the Absolute is inseparable, since each term illuminates the other and vice versa. The structure of consciousness regulates the phenomenon and enables the passage from phenomenal knowledge to knowledge in itself. In its ambiguity, arising from its dual reference both to what it posits as the in-itself and to what for it constitutes the knowledge of the in-itself, it expresses the dialectical nature of the Absolute, its reversibility, and the "reversal of roles between the measured and the standard of measurement"[84]. Only the experience of consciousness "can make way for a dialectic, because it alone is an openness,"[85] to something transcendental, and because the "relation to a transcendent"[86] is at the same time a relation to the self. Therefore: simultaneous reference of consciousness – in its *Zweideutigkeit*-to both, "this is to say that latent intentionality calls for reversals and not dependence in the unitary sense direction® of a noema in relation to noesis."[87] It is thanks to this reversibility – which Merleau-Ponty in The Visible and the Invisible defined as "ultimate truth"[88] – that knowledge is gradually restructured along with the transformation of the object[89] – according to that "solidarity" of activity and passivity that reflective philosophies have never been able to conceive – and truth, rather than shining in the coincidence and sterile immediacy of one with the other, shines through – ambiguously – precisely in their "exchange." Therefore,

> thus, it seems that dialectic (if it is defined entirely in this way) could be, essentially, phenomenology (in the Hegelian sense), that is, the relation to the self in relation to a transcendent [object] and vice versa. This dialectic is an intertwining of subject

84 Therein, p. 22.
85 Therein, p. 26.
86 Ibid.
87 Ibid.
88 Merleau-Ponty, M., *The Visible and the Invisible*, quot., p.155.
89 See Merleau-Ponty, M., *Philosophy and Non-Philosophy since Hegel*, quot..

> and object, because they are abstract moments of an 'experience'. Since the dialectical movement is a movement of content, it is not 'given' by consciousness; it exists only as content in relation to someone who experiences and lives it. No dialectical movement is possible without the duality: relation to the thing and relation to the Self. [...] Philosophy – that is, access to the absolute – appears to be essentially experience. Philosophy is an entering into phenomena in order to take part in their maturation and in experience. This is its character because in the relation of *experiri* with existing things, we can participate in the advent of knowledge.[90]

Thus, for Merleau-Ponty, dialectics as the self-movement of consciousness, that is, as the intentional relationship of a content with someone who experiences it, is essentially phenomenology, experience, an "itinerary in the course of which it learns, even though that which it learns was already there, in itself, before it, and even though it is nothing other than its passage to being-for-itself"[91]. There can be no dialectical movement without the dual relation to the thing and to the Self. This is why experience is the being-present-to-itself of consciousness, and philosophy can be conceived as an "entry" into phenomena and as access to the Absolute. In other words: it is the overcoming that must be considered as conservation; for the truth, from the very beginning, is already there; the beginning coincides with the end or with the result; there is, simultaneously, an "archaeological and teleological protension." Consequently, "to enter into the profundity of 'experience' [*Erfahrung*],"[92] of knowledge, means to grasp in the "new ontological milieu, which is the 'experience'"[93], The reciprocal metamorphosis of the in-itself and the knowledge of the in-itself, in which "the identity of savage consciousness

90 Therein, pp. 26-27.
91 Merleau-Ponty, M., *Themes from the Lectures at the Collège de France, 1952-1960*, quot.,
92 Id., *Philosophy and Non-Philosophy since Hegel*, quot., p. 46.
93 Therein, p. 91.

and reflected consciousness"[94] is given; it means to read "the self-movement of the preobjective and presubjective content"[95] in it as a discontinuous process, producing new meanings and at the same time fulfilling the past due to the necessity with which it is endowed.

In the commentary to the translation of par. 13, Merleau-Ponty further explains:

> It is not necessary to imagine consciousness either as a relation to the pure exterior, or as a pure relation to the self, or as the summation of the two (consciousness of the true plus consciousness of the knowledge of the true). Consciousness is the third and unique aspect for which the first two (the true and knowledge of the true) are distinguished by abstraction. This means that there is no true and no knowledge of the true. There is only one single term, and a second which is the same, and a third term, which is: openness onto…, examination of the self in the presence of. […] An openness before that which was knowledge can become an object, and that which was true can be restored to the level of simple 'knowledge'. Conscious-ness must not be thought alternatively as a relation to the self and a relation to another, but as the locus of a single explosion which produces selfness and consciousness of something. Consciousness serves neither as positive nor as negative, but as negativity at work – thus one cannot truly speak of consciousness as either an entity or as a "négatité".[96]

In the *Erfahrung* of Hegelian consciousness, Merleau-Ponty thus discerns the features of that "vertical world" whose ontological significance he aimed to articulate in *The Visible and the Invisible*. Likewise, in the *Zweideutigkeit* of consciousness, he perceives the "good ambiguity" – which he had already previously defined as "spontaneity that accomplishes what seemed impossible, given the separate elements" – which, as latent intentionality, precisely

94 Therein, p. 36.
95 Therein, p. 40.
96 Therein, p. 37.

characterizes Erfahrung as dialectic. Similarly, in the same *Bewusstsein*, he identifies the contours of a *fissure* (a crack) to be understood as "the locus of a single explosion which produces selfness and consciousness of something"[97]. In short, he discerns within the order of the phenomenon a new ontological domain, namely the *Erscheinung*, understood as the self-revealing "filigree" of the Absolute – that is, a dialectical movement of self-manifestation of the Absolute "by which going out of oneself is simultaneously returning into oneself, and vice versa," and which "gives rise to the segregation of interior and exterior as well as to their reversibility."[98] Dialectic thus emerges as the privileged domain that allows the French thinker to effect a "disruption" within traditional philosophy and to realize a non-philosophy grounded in a renewed recuperation of phenomenology. Dialectic as experience indeed implies a relation to a pre-objective, pre-categorial Being, which in many respects refers back to the antepredicative Being, to the wild Logos of which we are a part, from which we are *begriffen* – that is, "grasped," "taken," or "thought."[99] This is precisely why dialectic must be *Erfahrung* – that is, the "movement" of experience. Here, the chiasm constitutes the reversibility between the subjective and the objective, allowing all "movements" to maintain an openness and continuous relation among the various forms of consciousness, and consequently, also between experience and concept: "experience must be a 'concept', but 'concept' must be experience."[100] Thus, in this balance between experience and concept, which seems to exclude both skepticism and

---

97 Ibid.

98 Carbone, M. *Alla ricerca dell'*a-filosofia. *Merleau-Ponty* e la *Einleitung*, alla *Phänomenologie des Geistes*, quot., pp. 220-221 (my translation).

99 Merleau-Ponty, M., *Philosophy and Non-Philosophy since Hegel*, quot., pp. 39-40.

100 Therein, p. 43.

dogmatism, Hegel initiated a philosophy that could also be thought of as a non-philosophy.

The significance of this university course lies precisely in the fact that, within it, the complex relationship with Hegel becomes for the French philosopher an occasion for a confrontation with himself. Hegel, on the one hand, becomes Merleau-Ponty's privileged interlocutor for addressing his own philosophical questions, and, on the other hand, it seems almost as if the German philosopher himself provides Merleau-Ponty with the very sense of his questions, which the latter appears to have "directly" learned from him[101]. Thus, precisely in examining the aporias of Hegelian discourse, he comes to grips with the most pressing questions he himself had to confront.

The terms of the issue can be very briefly summarized as follows: on one hand, pure, "wild" experience – i.e., non-conceptual, un-conceptualized—becomes, in Hegel, a hallmark of skepticism; on the other hand, "comprehended," "unified" experience becomes an expression of dogmatism. The crucial point is identified by Merleau-Ponty when he asserts that

> philosophy was defined in 1807 as this double critique of experience by the concept and the concept by experience. We exclude impressionism and dogmatism. But these dangers exist to the second power: 1 One has recourse to experience as dogmatism (a judgment of history – the facts show that..., I did not say it, the things did. Consider, for example, the 1937 trials). 2 One has recourse to the concept as skepticism (I think that this experience is absolute – that I have attained the essence there – that 1937 [is] history in its nakedness). [...] Phenomenology and logic represent these two dangers. There is no solution as long as we remain at the antithesis: subject – object or consciousness of the external world – consciousness of the self. Hegel has surpassed this antithesis for a moment in his analysis

101 Lefort, C., *Philosophie et non-philosophie*, quot., p. 111 (my translation).

> of the Third Term. But he re-establishes the 'contribution' of philosophy. Therefore, his very recourse to the phenomenon is the most violent dogmatism. The others no longer know what they think. The philosopher understands them better than they understand themselves. In that respect, he aggravates the situation. His successors will protest against his dogmatism (Kierkegaard and Marx), and against his skepticism or his conciliation (Marx and Kierkegaard). [...] But these philosophers as well, in their opposition to Hegel, have the same oscillation between dogmatism and skepticism. The problem of a philosophy which might be non-philosophy remains *in toto* as long as one thinks consciousness or 'object' [*Gegenstand*].[102]

If, according to §§13 and 14 of the *Einleitung* to the *Phenomenology of Spirit* – explicitly examined by Merleau-Ponty in these course notes – Hegel's thought seemed capable of indicating to modern man the path to follow toward a "true philosophy," then considering §§15 and 16 (the latter including §17), Merleau-Ponty observes that Hegel's endeavor was, in fact, only a *conatus* culminating in skepticism; a skepticism that the Stuttgart philosopher was compelled to overcome, yet which did not prevent him from falling into the nets of dogmatism. This explains the subsequent critique by Marx, who, however, in turn – according to the French philosopher – "with his philosophy of the object falls back into the dogmatism of absolute subjectivity"[103].

To return to the impasse of skepticism versus dogmatism, it is clear that the misunderstanding thus created cannot be resolved either through bare experience or by appealing to "another source" (God, Reason, etc.), but requires "bringing mute experience back to the pure expression of its authentic meaning," as Merleau-Ponty had already emphasized in other works, such as *Signs*, *The Visible and the Invisible*, and *The Prose of the World*. That is to say, on the condition

102 Merleau-Ponty, M., *Philosophy and Non-Philosophy since Hegel*, quot., pp.43-44.

103 Therein, p. 309.

of abandoning consciousness as the ultimate reference, of no longer treating experience as a private property, and of restoring it as a primordial locus that entails its own reflexivity and development, so that knowledge of experience and experience of knowledge indefinitely refer to one another, and so that no absolute escapes history.

But when and why does Hegel's attempt to formulate a philosophy of non-philosophy fail? In the lecture of March 6, 1961, Merleau-Ponty writes: "The true: only that which turns out to be (work, becoming, experience) is absolute. There is no external relationship with the absolute, no instrumental means of joining it."[104]

And then, the crucial passage:

> No introduction to the absolute: we are there.[105]

Merleau-Ponty therefore reproaches Hegel for placing under the title of "Introduction" (*Einleitung*) the pages he was progressively examining and commenting on, which, in Merleau-Ponty's view, could not legitimately be considered – contrary to how they were later treated – as introductory to the entire system. For if Hegel speaks of the Absolute as *Erscheinung*, as *Offenbarung*, as *Erfahrung*, as being-present-to-itself, and if it is thus configured as phenomenology – that is, as "self-presentation of mind, an appearance which is not an effect of the absolute, but the absolute itself." – it is inadmissible that he should then distort and betray this meaning by treating all of it as merely an "introduction to".

> [...] the fragment which is translated without a title is subsequently called the 'Introduction.' Hegel himself no longer holds the view that there is no 'introduction' to the phenomenology of mind, which is the presence of the absolute. In 1817, at the time of the Encyclopedia, phenomenology again becomes a dis-

104 Therein, p.44.
105 Ibid.

> cipline, i.e. a part of science. It can no longer be 'the whole of the system from a certain point of view' (Hyppolite). In 1807, a relationship of Ineinander, a concentric condition, a reciprocal envelopment, then gives way to the enveloping thought of the 'positively rational' – or of the speculative.[106]

To consider the "presence of the Absolute" as an "Introduction" to phenomenology is to conceive phenomenology merely as a simple pathway to veritable knowledge – a "discipline," indeed, "part of science" – and not as the historical unfolding of the Absolute itself. Thus, in *Phenomenology of Spirit*, Hegel's principal aim – the circularity and reciprocity between experience and concept, nature and spirit, the measurer and the measured (a circularity and reciprocity that could have prevented the philosopher from lapsing into both metaphysical dogmatism and skeptical relativism) – is ultimately missed: "the failure to reunite philosophy and non-philosophy, which he wanted in the first place, is due to a domination of the philosophy of the concept over a philosophy of 'experience' (Capital)."[107]

The appearing of knowledge within experience and its becoming absolute knowledge occurs through *Sinngebung*, and this entails that the Absolute empties itself, becoming "indifference," "pure conservation," "because its tie with the vertical world of experience is released."[108]

Ultimately, it is a

> skeptico-dogmatism in the name of a nominal absolute. The absolute is practically the empty 'nothing' of skeptics, and the world is practically the positivity of all that 'endures' while in 1807 the negative was at work. I have said that the movement of experience to things spoken is inevitable. But this means that in a sense philosophy negates itself in formulating itself (what Kierkegaard will call Pharisaism); at least this occurs when phi-

106 Therein, pp. 49-50.
107 Therein, p. 58.
108 Therein, p. 50.

> losophy converts itself into contemplated significations. – The dissociation: 'the enduring' – the empty absolute, the disjunction in the order of phenomena was inevitable with a philosophy which remained a philosophy of consciousness, of representation and of the 'subject'. – Hegel admirably deepened these notions, made them supple, showing the paradoxical relationships between consciousness and the object and its metamorphoses. But in conserving the relation to the Self and the relation to the external world, he could not avoid the fact that the double envelopment was equivocal, and that the fissure appears at the level which phenomena must seal. Perhaps it is here that Hegel has not succeeded in his desire to link philosophy and non-philosophy.[109]

If the transition from experience to what is said requires "a *Zutat* of philosophy,"[110] in that "experience is necessary but not sufficient" to access truth, and "it must be comprehended" from the perspective of the philosopher who introduces an addition consisting in a "reversal of consciousness," then this implies that the philosopher, far from merely engaging in "pure observation," adds to the plane of experience that of truth (which is, in fact, his own truth) of that experience. Consequently, this conception not only perpetuates the separation between philosophy and non-philosophy but, in fact – as Merleau-Ponty himself seeks to explain – leads to a dogmatism even more violent and dangerous, precisely because it hides behind the supposed necessity of the truth of experience. The "reversal" (*Umkehrung*) of consciousness does not occur spontaneously thanks to the ambiguous (*zweideutige*) structure of consciousness itself but is carried out critically (by "conversion") by the philosopher: "the others no longer know what they think. The philosopher understands them better than they understand themselves."[111] But philosophy, understood in this way, becomes once again "Denken of the overview,"[112] a

109 Therein, pp. 50-51.
110 Therein, p. 308.
111 Therein, p. 43.
112 Therein, p. 72.

totalizing and all-encompassing thought. Merleau-Ponty thus identifies in Hegel that overview thinking which, already in The Visible and the Invisible, constituted the focal point of his critique of reflective philosophy, and he reiterates that our fleshly involvement with the world prevents us from thinking in terms of the "pure spectator," the geometrical or panoramic gaze. On the contrary, it emphatically expresses the character of reversibility. This reversibility manifests itself only in "a knowledge in which subject and object, savage consciousness and reflected consciousness reciprocate each other. Both fall within knowledge, which is therefore not our *Sinngebung*, itself."[113] It is therefore not possible to turn to the *unsere Zutat*, to the philosopher's intervention, to reconcile un-reflected life and reflection in order to grasp the authentic unfolding of the thing itself. The ambiguity of consciousness – which is co-originarity, reversibility, and continuous exchange between "wild" consciousness and reflected consciousness – is, in fact, the true passage to truth. This truth, consequently, is always partial and never a retrospective or additive truth imposed upon experience. Hence, it is apprehended in the mode of interrogation and in the sign of ambiguity. Philosophy, if it aspires to be genuinely questioning reflection, must renounce the "bad ambiguity" of the overview *Denken*, of exhaustive and "calculated" possession of the thing, because in so doing, "wanting to be all, it does not inhabit the things it discusses," and of which it presumes mastery, and thus, in reality, it lacks everything. Philosophy must therefore relinquish the claim to dwell at a distance, to see inside, to contemplate from without. According to Merleau-Ponty, the task is precisely to succeed in what Hegel, although certainly pointing toward a possible path, failed to accomplish: to create a concrete philosophy.[114] For the French philosopher, true and concrete philosophy is precisely that which grasps the meaning of human action

113 Therein, p. 306.
114 Cf. therein, p. 73.

– not simply given by pure consciousness or by some external objectivity, but formed, as Husserl had said, within intersubjectivity – or, equivalently, within intercorporeity. A crucial point is therefore the openness to *something*, to "that which is nothing," to the *il y a*. In other words, philosophy has both the task and the privilege of positioning itself as a pursuit of Being and of *Ineinander*. The fundamental problem, in truth, is to reconceive the "proximity and distance of philosophy," it's being "everywhere and nowhere," on the condition of granting consciousness – particularly self-consciousness – the power to hold within itself its own contradiction, to be chez-soi precisely in what is its opposite. It is therefore necessary not to fabricate "an illusory all-powerfulness – a negativity that is so total that it founds and digests everything and nothing, (…) in the name of 'knowledge.'"[115] Negativity needs to descend "into the flesh of the world"[116].

§ 3. *Hyper-dialectics and Inquiry*

The questions addressed by Merleau-Ponty in his final university course, discussed in the previous chapter, as well as those that remain implicit in many pages of *The Visible and the Invisible*, which we shall now examine, aim to outline a genuine "reformation" of the Hegelian dialectic. Dialectic, observes Merleau-Ponty, has failed its very essence because it has been articulated in univocal meanings; once formulated in theses and solidified into a closed system of laws governing the becoming of reality, it has – in fact – become "denatured," betraying the polymorphism of Being and its transcendence. It has reinstated the fracture between gnoseology and ontology; no longer configuring itself as a thought "at work within

115 Therein, p. 73.
116 Therein, p. 79.

Being," it has become an "embalmed dialectic," a "bad dialectic."

> The bad dialectic is that which thinks it recomposes being by a thetic thought, by an assemblage of statements, by thesis, antithesis, and synthesis; the good dialectic is that which is conscious of the fact that every *thesis* is an idealization, that Being is not made up of idealizations or of things said, as the old logic believed, but of bound wholes where signification never is except in tendency, where the inertia of the content never permits the defining of one term as positive, another term as negative, and still less a third term as absolute suppression of the negative by itself.[117]

If dialectic is understood as a *method*, as a principle of reason that penetrates the very fabric of the Absolute and identifies itself with it, it must immediately be recognized as a true degeneration into a legitimizing ideology: for, by imposing an external law upon things, it not only conceals them but also forecloses our access to them.

There are passages in *The Visible and the Invisible* that speak of a "second-order reflection," of a "return to," of a rebound upon… But *hyper-reflection* is never a methodological reflection, nor a reflection upon the conditions of its own possibility, nor an opening onto the immediate, the originary, or Being *tout court*, nor onto the transcendence, the silence of things, or the ontological difference. Rather – according to Merleau-Ponty's logic and in keeping with his celebrated definition – hyper-reflection is *all of these inseparably and simultaneously.*

This definition of hyper-reflection corresponds perfectly to that of *hyper-dialectic*, which gathers within itself the plurality of relations and what the philosopher termed *ambiguity*. Hyper-reflection, then, will be reflection upon reflexivity

117 Merleau-Ponty, M., *The Visible and the Invisible*, quot., p. 94.

itself—but we must not understand *hyper* as something that stands beyond or above; rather, it is a reflection upon that which makes reflection possible, without belonging either to reflection itself or to its opposite, the immediate. It unfolds within the very relations that constitute the possibility of ambiguity itself. [...] *Hyper-reflection* is thus, simultaneously, a reflection *in*, *of* (arising from, produced by), and *upon*... the intertwining, the ambiguity, the reversibility"[118].

Hyper-reflection, inseparably bound to the concept of *hyperdialectics*, here clarifies the relation between Being and beings – a relation which, strictly speaking, cannot manifest itself outside of the Absolute, since it belongs to it; and which can occur only in the awareness of the provisional, contingent, and partial nature of historical events, within which it has always already taken root. The *I* and the world are now affirmed in their reciprocity and their inherence to the *chair* (flesh). In this context, *hyperdialectics* takes shape as the organ of an ontological phenomenology that, precisely through the concept of *flesh*, reaches the primordial rhythmic pulsation of Being, the truth of its reversibility. Yet *hyper-dialectics*, as "a thought that (...) is capable of reaching truth because it envisages without restriction the plurality of the relationships and what has been called ambiguity "[119], lives within the dialectical movement of Being without claiming to detach from it or to dominate it from above. It knows, therefore, that the history of the expression of Being – which it must rediscover and retrace – will not be a linear and ascending path, but one punctuated by limited and partial surpassings, "encumbered

---

118 Shällibaum, U., *La possibilité de la réflexion philosophique*, in VV.AA., *Merleau-Ponty. Le philosophe et son langage*, in "Recherche sur la philosophie et le langage", n. 15, 1993, p. 345 (my translation). See also : Chiesa P., *La sublimazione come cifra della reversibilità,* in "Chiasmi International", n. 11, 2009, pp. 157-170.

119 Merleau-Ponty, M., *The Visible and the Invisible*, quot., p. 94.

with survivals, saddled with deficits"[120]. Dialectical thought thus relates to Being – to which, moreover, it itself belongs – as to a "system of multiple entries," one that can never be contemplated by an external, totalizing gaze, but must always be traversed according to the modalities of the *Ineinander*.

This alone makes it possible to reach the primordial symbolism of the brute Being (*l'être brut*) without stiffening its plasticity, for it embraces within itself both the circular movement that marks the ever-returning event of the *dehiscence* of the flesh, and the movement of *reprise* that outlines that teleology of the *wild Logos* which is also its own archaeology – the very process of its articulation.

"The failure of the dialectic as thesis or 'dialectical philosophy' is the discovery of this intersubjectivity which is not perspectival but vertical, which is, extended into the past, existential eternity, savage mind *(esprit sauvage)*"[121].

To the failure of dialectical philosophy conceived as a *thesis*, the French philosopher opposes the essentially "progressive" and "incomplete" character – "by definition" – of "the passage from philosophy to the absolute, to the transcendental field, to the wild and 'vertical' being"[122]. And what "is to be understood not as an imperfection (a *Weltanschauung* philosophy, unhappy consciousness of the Encompassing) but as a philosophical *theme:* the incompleteness of the reduction ('biological reduction,' 'psychological reduction,' 'reduction to transcendental immanence,' and finally 'fundamental thought') is not an obstacle to the reduction, it is the reduction itself, the rediscovery of vertical being"[123].

Merleau-Ponty thus insists on the *passage to the vertical*, to the deep, pre-predicative dimension of Being – a dimension which, dialectically, cannot be brought to expression without

120 Therein, p. 95.
121 Therein, p. 175.
122 Therein, p. 178.
123 Ibid.

at the same time losing its intrinsic truth. The issue here is transparent: the paradox lies precisely in speaking of the *gap* from within the gap itself; in saying the divergence while knowing, simultaneously, that it eludes speech precisely insofar as it is such; in saying it and, at the same time, calling into question the very possibility of saying it; in saying it and, *simul*, contradicting it[124]. All this leads to what Merleau-Ponty calls the "good dialectic," that is, the dialectic that cannot be converted into "meaning, thesis, or stated thing," for otherwise – as we have already seen – it becomes "enunciation, *Satz*, absolute knowledge," and once again installs itself outside the *écart*, outside the philosophy of non-philosophy. "The only good dialectic is that which critiques itself and surpasses itself as a separate statement"; it is the one that speaks the *écart* without turning it into a "new positive," that says it and, in saying it, simultaneously revokes itself, "critiques" itself, since it can only ever be an enunciation internal to that which it seeks to name.

If, as Merleau-Ponty writes, "the only good dialectic is the hyperdialectics," how can one fail to see that it leads precisely to this radical paradox — one that philosophy itself may no longer be able to bear? The impasse into which Merleau-Ponty runs appears, indeed, to be insurmountable. Is silence, then, the only possible way?

> The philosopher speaks, but this is a weakness in him, and an inexplicable weakness: he should keep silent, coincide in silence, and rejoin in Being a philosophy that is there ready-made. But yet everything comes to pass as though he wished to put into words a certain silence he hearkens to within himself. His entire "work" is this absurd effort. He wrote in order to state his contact with Being; he did not state it, and could not state it, since it is silence. Then he recommences.... One has to believe, then, that

---

124 Cf. on the theme of "contradiction," Vitiello, V., *La voce riflessa. Logica ed etica della contraddizione*, Lanfranchi, Milan 1994; and Id., *Non dividere il sì dal no. Tra filosofia e letteratura*, Laterza, Rome–Bari 1996.

> language is not simply the contrary of the truth, of coincidence, that there is or could be a language of coincidence, a manner of making the things themselves speak – and this is what he seeks. It would be a language of which he would not be the organizer, words he would not assemble, that would combine through him by virtue of a natural intertwining of their meaning, through the occult trading of the metaphor – where what counts is no longer the manifest meaning of each word and of each image, but the lateral relations, the kinships that are implicated in their transfers and their exchanges. It is indeed a language of this sort that Bergson himself required for the philosopher. But we have to recognize the consequence: if language is not necessarily deceptive, truth is not coincidence, nor mute.[125]

This passage – surely one of Merleau-Ponty's most celebrated – condenses, in evocative tones, the ultimate outcome of his thought, in which all the themes progressively explored by the philosopher converge into an exemplary synthesis. Sini, referring precisely to this passage, comments: "Language and truth constitute the very cipher of philosophy; philosophy strives to capture the muted words murmured by Being. Merleau-Ponty's project was to begin anew the history of the word – or rather, to wrest the word from history by tracing it back to the great silence of the world, within and from whose flesh the word explodes. Our vision of the human will remain superficial so long as we do not return to this origin, so long as we do not rediscover, beneath the murmur of words, the primordial silence, so long as we do not describe the gesture that breaks this silence. For this reason," Sini continues, "Merleau-Ponty proposes to take into consideration the word before it is spoken – against the backdrop of the silence that precedes it, that never ceases to accompany it, and without which it would say nothing. We must make ourselves sensitive to those threads of silence with which the fabric of speech is interwoven. It is in and from silence that I, world, and language emerge in their unitary correlation: enigmatic

125 Merleau-Ponty, M., *The Visible and the Invisible*, quot., p. 125.

poles of a single linguistic or semiotic play, masks of the Same, approximated distances in which, as Peirce maintained, each is where it functions – that is, where it announces itself and echoes. The word of the philosopher – that word which, as Merleau-Ponty says, must *accompany* this eventual play – is thus an echo of the world: always distanced from its own point of explosion. For Being is never before me, but surrounds me and, in a certain sense, traverses me. [...] My vision of Being does not take place from elsewhere, but from the very heart of Being. Thus, speech speaks only from the silence of the world – that silence which it guards and carries within itself. And precisely for this reason, it cannot say it (since it speaks and resounds in an echo without origin, or whose origin is precisely the silence to be said, the silence of saying); and precisely for this reason, it always says it – endlessly renewing the rebound of the echo that brings forth the world and marks the distance"[126].

The word is, therefore, an opening onto the world, *Offenheit*, inasmuch as it is only within it that the world occurs and announces itself. Here, Merleau-Ponty proposes – rather than Hegelian speculative dialectics – the Socratic-Platonic *dialéghestai*, in which the reversibility proper to the hyperdialectic is emblematically manifested: a reversibility that mediates itself with itself and thereby corresponds to the question of Being, which, having interrogated us, calls upon us to interrogate in turn. The late Merleau-Ponty embraces the Plato of dialogue and directs his reflections toward a decisive deepening of the theme of the language of the world (*chair langagière*), in which silence and speech are inscribed along with their mutual deferral. In this sense, one can understand the philosopher's expression, according to which "Far from harboring the secret of the being of the world, language is itself a world, itself a being – a world and a being to the

126 Cf. Sini, C. *Il silenzio e la parola*, quot., pp. 10-13 (my translation).

second power, since it does not speak in a vacuum, since it speaks *of* being and *of* the world and therefore redoubles their enigma instead of dissipating it"[127]. Here, language, as such, expresses inseparably both the world made speakable as word and the world that always withdraws from it, retreating into an "infantile" silence. The ambiguity, the doubleness/sameness of language as silent word: "Language arises from... and speaks... all the silences; it articulates what the world, Being, the sensible, and silence have always meant to convey". "The relation between the thematization and the behavior is a dialectical relation: language realizes, by breaking the silence, what the silence wished and did not obtain"[128].

> Speech (...) brings a ferment of transformation that will give the operative signification. Just as the flesh (or gaze) emerged at the crossing of the visible and the invisible and was nothing but the gap, the *in-between*, so too does speech arise at the point where language and silence intersect. The second term is the depth, the thickness of the first; it appears only within it and continues to envelop it. Silence is not absolute, just as language is not absolute: the former teems with words – there are voices of silence – while the latter is woven with silences. One can never say everything, never exhaust silence; for this reason, the effort of expression is always at its first day, begins anew with each human being, and will never be completed. Silence seeks words, and words seek silence. Philosophy is language, but it consists in rediscovering silence.[129]

However, "what will – asks Merleau-Ponty – this silence be? [....] this silence will *not be the contrary* of language [....]: This reversal itself – *circulus vitiosus deus* – is not hesitation, bad faith and bad dialectic, but return to Sigh′, the abyss"[130].

---

127 Merleau-Ponty, M., *The Visible and the Invisible*, quot., p. 96.
128 Therein, p. 176.
129 Therein, p. 228
130 Therein, p.179.

He continuous cycle of rejection and retrieval of silence signifies that there are two silences between which the word is situated – diacritically – whereby, according to Lefort, "it gives expression to an experience that is mute and ignorant of its own meaning, but only in order to make that experience appear in its purity; it does not break our contact with the things, but it draws us all things only in order to awaken us to the truth of their presence and to render palpable their relief and the tie that binds us to them. At least such is the speech that speaks in conformity with its essence and, where philosophical discourse is concerned, that does not cede to the vertigo of eloquence, does not wish to suffice to itself or close in upon itself and upon its sense, but opens upon and leads to the outside. But if speech, which is born from silence, can seek its conclusion in silence and make that silence not be its contrary, this is because between experience and language there is, in principle, exchange; it is because experience is not something one could coincide with, because it bears a transcendence, since already, in itself, it is differentiation, articulation, structuration, and because in some way it calls for language; it is because language is also experience, because there is, [....] a *being* of language in which the enigma of being is repeated, because beyond the movement of the pure significations there remains the silent mass of the discourse, that which is not of the order of the sayable, and because the greatest merit of expression is to disclose this continuous passage from the word to being and from being to the word, or this double openness of the one upon the other"[131]. That is why, if on the one hand philosophy, being language, "consists in rediscovering silence," on the other hand, however, precisely in order for this to be possible, we must "start speech anew and not give up on it."[132]

131 C. Lefort, *Foreword* to Merleau-Ponty, M., *The Visible and the Invisible*, quot., pp. XXVIII-XXIX.

132 Merleau-Ponty, M., *The Visible and the Invisible,* quot., p. 213.

Precisely "to think through this exchange" was the task to which Merleau-Ponty felt compelled and urgently drawn in *The Visible and the Invisible*. It is certainly significant that the philosopher's final words in this work stand as a tangible mark of this imperative to think:

> […] the whole landscape is overrun with words as with an invasion, it is henceforth but a variant of speech before our eyes, and to speak of its "style" is in our view to form a metaphor. In a sense the whole of philosophy, as Husserl says, consists in restoring a power to signify, a birth of meaning, or a wild meaning, an expression of experience by experience, which in particular clarifies the special domain of language. And in a sense, as Valery said, language is everything, since it is the voice of no one, since it is the very voice of the things, the waves, and the forests. And what we have to understand is that there is no dialectical reversal from one of these views to the other; we do not have to reassemble them into a synthesis: they are two aspects of the reversibility which is the ultimate truth.[133]

We can also speak of a reversibility of language between thought and word, because they anticipate each other and constantly replace each other: "Operative language makes us think, and living thought magically finds its words". Indeed, "There is not thought and language [….], Expressive operations take place between thinking language and speaking thought; not, as we thoughtlessly say, between thought and language"[134]. Thus, the word becomes signifying and

---

See at this point: Wakdenfels B., *Faire voir par les mots. Merleau-Ponty et le tournant linguistique,* in "Chiasmi International", n. 1, 1999, pp. 57-63; De Saint Ubert E., *Conscience et expression chez Merleau-Ponty: l'apport du cours inédit sur le monde sensible et le monde de l'expression,* in "Chiasmi International", n. 10, 2008, pp. 85-107; Leoni F., *L'esperienza della filosofia. Senso e significato della parola filosofica nell'ultimo Merleau-Ponty,* in "Chiasmi International", n. 10, 2008, pp.183-193.

133 Therein, p. 155.

134 Id., *Signs*, quot., p. 18.

foundational in the sense that it can no longer be considered a mere "statement" (Satz); rather, it must be a thinking word, without reference to a *Sachverhalt*, word and not language. This is because – Merleau-Ponty emphasizes – "It is indeed speaking that constitutes, *in front of* myself as a signification and a subject of signification, a milieu of communication, an intersubjective diacritical system which is the spoken tongue *[la langue*] in the present, not a "human" universe, an objective spirit)"[135].

Moreover, in *The Visible and the Invisible*, we can read:

> Being opens itself to me in a *logos*, and the 'natural magic' of the word opens me to this *logos* of Being. […] there is an essence beneath us, a common nervure of the signifying and the signified, adherence in and reversibility of one another […]. Speech prolongs into the invisible, extends unto the semantic operations the belongingness of the body to being and the corporeal relevance of every being. […] As the world is behind my body, the operative essence is behind the operative speech also, the speech that possesses the signification less than it is possessed by it, that does not speak *of it,* but speaks *it,* or speaks *according to it,* or lets it speak and be spoken within me, breaks through my present[136]

The word, now intertwined with the flesh, thus appears in an explicit ontological dimension: the *chair langagière*. The Merleau-Pontian perspective of hyperdialectics as dialogical reveals its full significance precisely in light of this new and decisive reflection on language, a privileged access to the problematics of Being.

Language becomes, in fact, a modality of Being according to the ontology of flesh and of the chiasm. In this sense, the dialectic without synthesis between myself and the world, structured by the chiasm, finds its parallel in the essential reversibility between myself – who, in responding to the

135 Id., *The Visible and the Invisible*, quot., p. 175.
136 Therein, p. 118.

call of things, operates unconsciously within language – and language (*langagier*), which materializes in the direction of meaning. Thus, on this side of conventional means of expression – which manifest to the other my thought, insofar as meanings are already given in me as in him for every sign – one must recognize, according to Merleau-Ponty, a primordial operation of signification that occurs already at the level of the lifeworld (*monde-de-la-vie*), thanks to transcendental intersubjectivity. Against every philosophy of deduction and construction, the French thinker once again privileges the antepredicative, as the appropriation of the sense of the world, open stratigraphically toward successive explication:

> We live in a world where speech is an *institution*. For all these many commonplace utterances, we possess within ourselves ready-made meanings. They arouse in us only second order thoughts; these in turn are translated into other words which demand from us no real effort of expression and will demand from our hearers no effort of comprehension. Thus, language and the understanding of language apparently raise no problems. The linguistic and intersubjective world no longer surprises us, we no longer distinguish it from the world itself, and it is within a world already spoken and speaking that we think. We become unaware of the contingent element in expression and communication, whether it be in the child learning to speak, or in the writer saying and thinking something for the first time, in short, in all who transform a certain kind of silence into speech. It is, however, quite clear that constituted speech, as it operates in daily life, assumes that the decisive step of expression has been taken. Our view of man will remain superficial so long as we fail to go back to that origin, so long as we fail to find, beneath the chatter of words, the primordial silence, and as long as we do not describe the action which breaks this silence.[137]

From this point, two thematic levels unfold: the first concerns the question of "thought as second word," while the

137 Id., *Phenomenology of Perception*, quot., pp. 213-214.

second, grounded in the first, refers to the "primacy of the question."

In what sense, then, does Merleau-Ponty speak of thought as a second word? And further: is it only thought as such that is considered "second," or is this rather a dimension that involves us already, from the start, as *existent* beings prior to being *thinking ones*?

Thought, the philosopher explains, is a secondary and empirical operation of translation, of codification and decoding, a mere technical relation of a sound and a sense that are joined only by convention. It is, therefore, reflection upon the un-reflected, expression of an expressible, a linguistic construction of and on the world. Yet Merleau-Ponty goes far beyond this clarification – or rather, he is compelled to carry this discourse to its deepest consequences. Indeed, it is not simply a matter of recognizing thought as "second word" – which may in fact be plausible and even acceptable – but rather of questioning, *radicitus*, the fact that we ourselves are second word… that we ourselves are language of language, word of word, "second word," in this very sense.

How does the philosopher arrive at these conclusions? And what consequences follow from them?

We have seen that, in *The Visible and the Invisible*, Merleau-Ponty introduces language as a modality of Being, as a spontaneous power to express the existent, as a privileged realm capable of opening the ontological dimension of flesh, that is, the *chair langagière*.

Yet this power of language to express the existent remains, despite everything, obscure and invisible:

> Behind the spoken *cogito*, the one which is converted into discourse and into essential truth, there lies a tacit *cogito*, myself experienced by myself. But this subjectivity, indeclinable, has upon itself and upon the world only a precarious hold. It does not constitute the world, it divines the world's presence round about it as a field not provided by itself; nor does it constitute the word, but speaks as we sing when we are happy; nor again

> the meaning of the word, which instantaneously emerges for it in its dealing with the world and other men living in it, being at the intersection of many lines of behavior, and being, even once 'acquired', as precise and yet as indefinable as the significance of a gesture. The tacit *cogito*, the presence of oneself to oneself, being no less than existence, is anterior to any philosophy, and knows itself only in those extreme situations in which it is under threat: for example, in the dread of death or of another's gaze upon me. […] The tacit *cogito* is a *cogito* only when it has found expression for itself.[138]

Thus, while the spoken cogito is language, the tacit cogito is that un-reflected dimension which can become reflection only by taking the form of word: the movement by which the silent cogito comes to self-knowledge is the movement of linguistic signification. Indeed, Merleau-Ponty continues,

> if we were to make completely explicit the architectonics of the human body, its ontological framework, and how it sees itself and hears itself, we would see that the structure of its mute world is such that all the possibilities of language are already given in it. Already our existence as seers (that is, we said, as beings who turn the world back upon itself and who pass over to the other side, and who catch sight of one another, who see one another with eyes) and especially our existence as sonorous beings for others and for ourselves contain everything required for there to be speech from the one to the other, speech about the world.[139]

Certainly, the entire Merleau-Pontyan discourse analyzed thus far ultimately demonstrates how the problem of Being and that of language are mutually interpenetrating. From the ontological perspective delineated in *The Visible and the Invisible*, language is established definitively as a privileged access to the question of Being – not because Being could be reduced to formal predication, but because language allows us to grasp and express its structures, to come into contact with the

138 Therein, pp. 469-470.
139 Id., *The Visible and the Invisible*, quot., p.155.

intersubjective community, and thus to reach a universe of raw being and of co-existence. The error of various philosophies has consisted in defining language as a closed domain unto itself, as that which speaks exclusively of itself. In reality, however, "what is lived is lived-spoken" and "language is not a mask over Being, but – if one knows how to grasp it with all its roots and all its foliation – the most valuable witness to Being"[140]. Considering the speaking word, the intertwining of the visible and the lived with language, of language with the visible and the lived, the exchanges between the articulations of its silent landscape and those of its word, and finally, that operative language which does not need to be translated into significations or thoughts, that thing-language which brings to the surface all the profound relations of the lived experience in which it has taken shape, then, in this sense, Merleau-Ponty once again transcends the "world of pure consciousness" and its acquired and constructed articulations, posing the problem in its most radical terms and thereby arriving at an original perspective: a being that lies behind the verified-being, the raw spirit from whose horizon constructed objects and significations emerge.

The problem is thus brought back to its ontological matrix, to the dimensionality of being and to our relation with it. One then understands why philosophical reflection "itself is language"; and how "this does not disqualify it from speaking of language, nor from speaking of the pre-language and of the mute world which doubles them: on the contrary, philosophy is an operative language, that language that can be known only from within, through its exercise, is open upon the things, called forth by the and continues an effort of articulation which is the Being of every being."[141] This is because, as Lefort observes,

140 Therein, p.126. See interesting texte by Pierron J.P., *La prose du monde ou le monde comme un texte? Ricœur, lecteur de Merleau-Ponty,* in "Chiasmi International", n. 17, 2015, pp. 309-323.

141 Therein, pp.126-127.

"I engage with Being before knowing it; my questions emerge prior to becoming philosophical interrogation. Thus, I must bear in mind what exists before any thesis, before any reflection; I must maintain this engagement with Being amid the semi-silence and semi-darkness of consciousness."[142]

All this implies, quite naturally, that the method of accessing Being is "indirect," "secondary"; "*One cannot make a direct ontology.* My "indirect" method (being in the beings) is alone conformed with being (...)."[143]

To utter the silent language of Being, to produce thereby, in a sort of detonation, the innate knowledge of the body through which Being manifests itself, to gut, so to speak, the mystery and essential opacity of things – this is the demand that characterizes the indirect method of the French philosopher.

It is certainly visible, in the late Merleau-Ponty, a "strong Heideggerian temptation," insofar as philosophy is brought back to the "explication" and "recollection" of Being:

> The "amorphous" perceptual world that I spoke of in relation to painting – perpetual resources for the remaking of painting – which contains no mode of expression and which nonetheless calls them forth and requires all of them and which arouses again with each painter a new effort of expression – this perceptual world is at bottom Being in Heidegger's sense, which is more than all painting, than all speech, than every "attitude," and which, apprehended by philosophy in its universality, appears as containing everything that will ever be said [...]: it is the λόγος ενδιάθετος which calls for the λόγος προφορικός.[144]

The last sentence of the passage just quoted brings us back to the central question: we are the "second speech," therefore λόγος προφορικός manifestation of Being. And we are so because we are solicited, questioned by that λόγος ενδιάθετος,

---

142 Lefort, C. *L'idée d'Être brut et d'Esprit sauvage,* in "Les Temps Modernes", n. 184-185, 1961, p. 271 (my translation).

143 Merleau-Ponty, M., *The Visible and the Invisible*, quot., p.179.

144 Therein, p.170.

that spontaneous logos that speaks from the depths of Being. But then, from all that has been said so far, it becomes possible to fully grasp not only the intrinsically dialogical character of Merleau-Ponty's hyper-dialectic, but also the very sense of philosophical interrogation, which Merleau-Ponty now situates as both the archeological and teleological principle of our inscription in Being. The consequences of this path of thought unfold entirely within a fertile dialegesqai from which arises—as we have already anticipated – a conversation (*ein Gesprach*)[145] with Being, now characterized as the primacy of the question.

It is not the philosopher who – from the height of his position – interrogates Being but Being itself that asks to be interrogated; it is Being that calls us to respond to its inexhaustible question of meaning so that it may be. Being is thus interrogation, a continuous questioning and search for sense. Being is interrogation because it is constitutively absence, lack, and thus a demand to fill a gap. Yet this and the surface of the λόγος προφορικός, between the invisible and the visible, between silence and speech, reveals itself as a mere possibility.

What characterizes the ontological question in its possibility is the "turning back upon itself of the question" (*retourner sur elle-même*). "As soon as it is posed, this question in its second power could not be erased." The question is the only way to bring the interval to expression, because it articulates it and, in articulating it, defers it, distances it from itself: it asks what the interval is and, at the same time, what it is to ask. "Questioning" is "the sole manner of expressing this perpetual elusiveness."[146]

145 On this point, we must duly refer to Heidegger, M., 1973 and 1984, as the twentieth-century philosopher who most deeply engaged with the "question of language."

146 See Merleau-Ponty, M., *Themes from the Lectures at the Collège de France, 1952-1960* quot.

The questioning of philosophy is therefore not a form of agnosticism, nor is the question about philosophy a preliminary question, nor, finally, a way of characterizing philosophy as a heuristic ideal always oriented toward a new question. The question is "is the proper mode of our relationship with Being"[147]. It is a questioning addressed to Being "that would form itself *within Being*"[148], "endo-ontology"[149], a doubling of the question itself. "What there is to be grasped – writes Merleau-Ponty – is a dispossession"[150], an ascent that is never accomplished for it is "an ascent on the spot".[151] The task of the philosopher is therefore to respond – interrogatively – to the spontaneous, primordial question of raw Being; and here one can discern a task that is not exclusively of a theoretical order, but also of an ethical one, aimed at recovering – in the ante-predicative dimension – that dialogue which we ourselves are, thanks to which "we are making a philosophy of the *Lebenswelt", as well as* "our construction (in the mode of "logic") makes us rediscover this world of silence."[152] It is therefore a matter of traversing this silence, of experiencing it, of taking it into account and, in some sense, of producing it. The λόγος ενδιάθετος is not simply mute speech, silence that "solicits" speech (λόγος προφορικός), rather, it is speech that has always already been open to slippage into nothingness, for it is logos without Voice, speech that no longer rests upon any intention-to-say. To experience this language, then, properly means to put oneself at stake, to surrender to the greatest risk. Indeed, "To dwell in language without being summoned by any Voice, to simply die without being called by death, is perhaps the most abyssal experience; yet precisely for this reason, it is also, for humans, the most habitual experience, their ethos,

147 Id., *The Visible and the Invisible*, quot., p.129.
148 Therein, p.215.
149 Therein, p.226.
150 Therein, p.266.
151 Therein, p.177.
152 Therein, p.170.

their dwelling, which, in the history of metaphysics, always appears already demonically split into living and language, nature and culture, ethics and logic, and is therefore accessible only within the negative articulation of a Voice. And perhaps, only from the eclipse of the Voice, from the absence of a locus for language and the death of the Voice, does it become possible for humans to experience their own ethos in a way that is no longer merely sigetic"[153]. It is in this direction that Merleau-Ponty invites us to reflect, and it is perhaps for this reason that interrogation – in the sense of the taking place of a Voice, in the sense of responding to the primordial question – remains the final and the only *possibility of philosophy.*

"Language carries within itself its own end"[154], and this end is itself, it is the articulation or movement of Being itself; it is the word that becomes world from the depths of silence, from the mute world of the λόγος ενδιάθετος. This, however, does not mean that the word breaks the silence; on the contrary: paradoxically, the word is silence because it speaks. Precisely insofar as the word speaks, this speaking silences the dimension that it, structurally, cannot articulate. In this sense, Merleau-Ponty speaks simultaneously of a "speaking that is silence" and a "silence that is speaking," thereby reaffirming the meaning of "reversibility as ultimate truth." But if things are thus – if silence belongs to the word, if it is its very gap, its own distance – then this implies the impossibility of ever coinciding with Being. Hence, the paradox of expression, of the expressible, points directly to the paradox of Being itself, which cannot be grasped in a "sayable without remainder." Moreover, "the fissures that the event of truth reveals cannot be healed by anticipating, by going back, or by going beyond. The irrevocable 'shattering' of Being (éclatement *d'Être*), of

153 Agamben, G., *Il linguaggio e la morte*, Einaudi, Turin 1982, pp. 120-121 (my translation).

154 See Merleau-Ponty, M., *Themes from the Lectures at the Collège de France, 1952-1960* quot., (my translation).

which Merleau-Ponty speaks, rests on the fact that the very origin fragments and moves in different directions. If the sole origin shatters, pluralizes, and differentiates, then truth itself multiplies; it no longer has any stable locus and begins to wander." One then wonders whether even reversibility, that regressive movement without synthesis, which the philosopher considers as ultimate truth, does not make the act of doing and speaking spin excessively upon itself."[155]

In Merleau-Ponty, the awareness of this problem is very much alive, perhaps the most arduous to confront, precisely because it primarily engages the intrinsic meaning of indirect or interrogative ontology. In an unpublished working note, the philosopher writes on the matter:

> Perhaps one must begin directly from a stance of interrogative ontology, that is, from a description of Being in the present, founded upon the elimination of the ontic, of the causa sui, of the *ens necessarium*.
>
> Perhaps also: to start from the results of the *Phenomenology of Perception* and show that it must be transformed into ontology:
>
> 1. to move from the affirmation of the "perceived" to that of "raw Being";
> 2. to move from the idea of the body as subject to that of the being of indivision.
>
> To thematize interrogative ontology as a vision of the framework nature–human–Being.
>
> To say that the conception of Philosophy presented here depends on what will be said later about language—and does not simply determine it. Likewise, this ontology depends on what

155 Waldenfels, B., *Vérité à faire. La questione della verità in Merleau-Ponty*, in VV.AA., *La prosa del mondo. Omaggio a Merleau-Ponty,* Quattroventi, Urbino 1991, p. 90; cf. also Id., *Twovards an Open Dialectics*, in "Dialectics and Humanism", n. 3, 1976, pp. 91-101; and Id., *Faire voir par les mots. Merleau-Ponty et le tournant linguistique*, in "Chiasmi International", n. 1, 1999, pp. 57-63. See also: Simms E.M., *Egocentric language and the upheaval of speech. A Merleau-Ponty inspired study of language acquisition,* in "Chiasmi International", n. 12, 2010, pp. 287-309.

> will be said about things, life, and *physis*, and does not simply determine them.
>
> Circularity.[156]

It becomes clear, from these reflections, that Merleau-Ponty's project increasingly concerns the idea of a circular philosophy, unfolding as a sort of recommencement of the history of the word; and this precisely because it is there, in this attempt to continually recount the expressible, that the very sense of philosophizing is at stake. The world, in its being world, becomes word; it is the disclosure of that metaphor which both sustains and surpasses it. Every word is, in fact, woven with a metaphorical sense, a sense that therefore extends beyond and across itself. The ambiguity lies precisely in this dialectical-dialogical movement, whereby "there is, as it were, a spirit of language, and at the same time, the spirit is always full of language."[157] Following this essential ambiguity of language, we can fully grasp the meaning – not solely phenomenological, but also ontological – of Merleau-Ponty's expression "the word is at hand". This expression, in fact, embodies the tacit awareness that not only is speaking a way of dwelling in the world, but – in a deeper sense – it is the recognition that we are already in the world even before we articulate a word. Merleau-Ponty therefore writes that "Man can only make himself present to the world and to others through language, [because] language is, in everyone, the central function that constructs a life as a work, and that transforms even our difficulties of being into motives for life"[158].

It is evident that the French philosopher seeks, through language, the constitutive and originary dimension of Being, so that – in the end – it is language itself that projects silence

156 Merleau-Ponty, M., *Nota inedita*, Boîte V, quot., (my translation).
157 See Id., *Themes from the Lectures at the Collège de France, 1952-1960*, quot., (my translation).
158 Therein, (my translation).

behind itself, and silence contains, maintains within itself, language as its own excess, as *esprit langagier*, as *Offenheit*. "For Merleau-Ponty, it is Being that speaks in us, and not we who speak of Being. And philosophy, precisely as 'Being speaking in us,' a mute expression for itself, is creation. But it is a creation that is, at the same time, a reintegration of Being (since it seeks to recover its origin). It is thus creation in a radical sense: creation that is simultaneously adherence. Art and philosophy are contact with Being precisely as creations. Being is that which demands of us creation so that we may have its experience."[159]

In this context, Merleau-Ponty writes about Paul Klee's painting:

> It is impossible to say that nature ends here and that man or expression starts here. It is, therefore, mute Being which itself comes to show forth its own meaning. Herein lies the reason why the dilemma between figurative and nonfigurative art is badly posed; it is true and uncontradictory that no grape was ever what it is in the most figurative painting and that no painting, no matter how abstract, can get away from Being, that even Caravaggio's grape is the grape itself. This precession of what is upon what one sees and makes seen, of what one sees and makes seen upon what is – this is vision itself. And to give the ontological formula of painting we hardly need to force the painter's own words, Klee's words written at the age of thirty-seven and ultimately inscribed on his tomb: 'I cannot be caught in immanence'.[160]

These statements by Merleau-Ponty concerning Being clarify what the philosopher had already written in May 1960, in a working note entitled *The Philosophy of the Sensible as Literature*: "the world, Being, are polymorphism, mystery and

159 Philippe, M.-D., *Une philosophie de l'Être est-elle encore possible?*, quot., p. 131 (my translation)
160 Merleau-Ponty, M., *Eye and Mind*, in the Primacy of Perception, quot., p.188.

nowise a layer of flat entities or of the in itself"[161]. Firstly, it can be observed that the problem of the identity and difference of Being – of Being as "serpentine movement" – leads the philosopher to speak of an *ouverture* of the *Umwelt* that is modulated according to language. For, "language is precisely the system of differentiations through which the subject's relation to the world articulates."[162] The *langagier* Being contains – or rather, is – the interchange, the reversibility of myself and the world; it is, in this sense, a chiasmatic figure, and thus embodies ambiguity and transcendence. Being thus conceived is the ultimate ground of every essence and every existence; it is the primordial soil of predicative being.

The pursuit of the origin, of the beginning of Being, has, ultimately, been the guiding thread of Merleau-Ponty's philosophical reflection, especially since *Phenomenology of Perception*, where one already perceives the French philosopher's need to grasp the sense of Being prior to and beyond any objectifying thought, and thus beyond the vantage of detached overview. This need would later culminate in a decisive philosophy of interrogation, aimed at definitively closing the bridges with traditional philosophy. What, according to Merleau-Ponty, can bear the full weight of this new mode of thinking Being is "interrogative ontology," since only it returns upon itself, only it doubles itself as it positions itself, only it is ever anew in disjunction from itself, and only it reveals – broadly speaking – a question raised to the second power.

Thus, "far from consisting in retrieving in Being a specular positivity, the work of philosophy is interrogation into its own possibility. [...] This is therefore the stake in relation to this long critique of reflective philosophies: to discover interrogation as the ultimate relation to Being and as an

161 Id., *The Visible and the Invisible*, quot., p. 252.

162 See Id., *Themes from the Lectures at the Collège de France, 1952-1960*, quot., (my translation).

ontological organ; to discover, consequently, that the existing world exists according to the interrogative mode. As an object of interrogation, Being is not alien to the order of sense: the characteristic of a question is certainly that it entails the possibility of an answer, at least as a horizon. [...] But the characteristic of interrogation as pure interrogation is that the question accepts no answer capable of concluding it."[163]

The sense, then, is precisely an interrogative sense, always present as questioning, always *derrière*, almost retreating in relation to the question that reaches it. This interrogative sense thus manifests simultaneously as both the answer and the renewal of the question within the answer itself, which implies that it is present only at a distance: it is, in other words, solicitation, a call rather than positivity.

Here emerges the dual polarity of interrogation, the doubling of the chiasm: Being is that which I interrogate, that to which my questioning is directed; but it is equally – and at the same time – that which interrogates me and awaits from me a response that will further extend the question. In this sense, interrogation defines the ultimate and primordial reality of an openness to Being, of which subject and object are already abstract moments in themselves: since philosophy is recognized as interrogation, one must certainly say that it is precisely Being that interrogates itself within it, manifesting as a question upon and about itself.

Defining philosophy as interrogation is therefore to recover interrogation as the mode of Being of Being itself. Indeed – as Merleau-Ponty explains – as soon as interrogation is subordinated to the subject, it is immediately reabsorbed into pure ideality and resolved within representation; whereas it is necessary to understand interrogation as the very being of the subject, to reintegrate the subject into Being as its own teleology.

163 Barbaras, R. *De l'être du phénomène. Sur l'ontologie de Merleau-Ponty*, Millon, Grenoble 1991, p. 166 (my translation).

There is, then, a depth of Being that does not rest on its distance from the subject – for that would once again reinstate a thought of coincidence – but it is Being itself that maintains itself in its own distance; it is Being that keeps itself at a distance from itself. Once again, and in a more radical form, the French philosopher expounds his principle of the non-coincidence of Being with Being:

> What we propose here, and oppose to the search for the essence, is not the return to the immediate, the coincidence, the effective fusion with the existent, the search for an original integrity, for a secret lost and to be rediscovered, which would nullify our questions and even reprehend our language. If coincidence is lost, this is no accident; if Being is hidden, this is itself a characteristic of Being, and no disclosure will make us comprehend it. A lost immediate, arduous to restore, will, if we do restore it, bear within itself the sediment of the critical procedures through which we will have found it anew; it will therefore not be the immediate. If it is to be the immediate, if it is to retain no trace of the operations through which we approach it, if it is Being itself, this means that there is no route from us to it and that it is inaccessible by principle [...] The "answer" is higher than the "facts," lower than the "essences," in the wild Being where they were, and – behind or beneath the cleavages of our acquired culture – continue to be, undivided.[164]

To think the Being from the standpoint of interrogation also entails renouncing the distinction of "layers" or "levels" within it. It makes no sense to oppose, as Merleau-Ponty still did in the period of the *Phenomenology of Perception*, the order of the properly perceived to that of the thought. Perception, vision, are already interrogation: they do not open onto a *thing-in-itself*, but already disclose a sense; they are certainly a thought, yet in its raw or "wild" state. Sense emerges in them only as the very depth of the world; thought arises in them only as solicitation, as an appeal.

164 Merleau-Ponty, M., *The Visible and the Invisible*, quot., pp.121-122

In this, pure visibility is already an answer, carrying within its depth every response yet to come. But, correlatively, language itself also constitutes interrogation, an "inspired exegesis," for this silent *logos*, which is vision, does not lighten itself of its depth in favor of a pure ideality, but is instead brought back into it. Then, Merleau-Ponty writes: "When philosophy ceases to be doubt in order to make itself disclosure, explication, the field it opens to itself is indeed made up of significations or of essences – since it has detached itself from the facts and the beings – but these significations or essences do not suffice to themselves, they overtly refer to our acts of ideation which have lifted them from a brute being, wherein we must find again in their wild state what answers to our essences and our significations."[165]

In other words, it means "to return to this idea of proximity through distance, of intuition as auscultation or palpation in depth, of a view which is a view of self, a torsion of self upon self, and which calls 'coincidence' in question."[166] The meaning of Merleau-Ponty's project becomes still clearer: it is to reaffirm decisively the importance of a philosophical reflection grounded in a dialectical conception of Being which – far from constituting itself as a completed and final synthesis within a transparent horizon of meaning – establishes itself instead as a *"double reference,"* a *"covering-over,"* and a *"surpassing"; as* come "the identity of the retiring into oneself with the leaving of oneself, of the lived through with the distance."[167]

On this basis, one can also understand the meaning – in many respects cryptic – of an unpublished working note dated October 6, 1958, in which Merleau-Ponty writes that "in order to have a philosophy with multiple openings, it needs have a philosophy with a single access point: the multiple will truly

165 Therein, p. 110.
166 Therein, p. 128.
167 Therein, p. 124.

be such only if each of the openings presents itself as the only one."[168]. The openings of Being (*ouvertures d'Être*) to which the French philosopher refers are thus the possible openings of the *Umwelt* – openings that not only cannot be conceived within the order of "human representation," but that constitute the ontological dimension wherein "it is not we who perceive, [but] it is the thing that perceives itself over there; it is not we who speak, [but] it is the truth that speaks itself in the depths of speech."

All this implies that philosophy itself cannot be regarded as the sole avenue of access to Being – for such a claim would amount to a relapse into the Western conception of the history of thought – but must instead open itself to other, "non-philosophical" ways (hence the meaning of Merleau-Ponty's proposal of a *philosophy of non-philosophy*) of responding to Being.

The "Being of interrogation," as the French philosopher now defines it, therefore designates *mediation* as the ultimate reality (and here the full influence and weight of Hegelian reflection are evident), and, in a certain sense, there is nothing but mediation.

> It is not merely a matter of an actual interaction that would still allow for the distinction of essences. One must not think of the world as a particular case of something general that would be studied by logic: logic is the study of a *mode of being – the* lekton – and ontology cannot be subjected to logic. What is at stake is understanding how the modes of being articulate themselves within Being.
>
> Such an inquiry is rendered necessary by the crisis of all our categories — a crisis testified to by the very exhaustion of each of them. For example: the crisis of the notions of *nature*, *man*, and *God*, each of which passes over into the others. This confusion reveals their connection, their solidarity within Being, and thus the necessity of a radical inquiry into the *articulation of Being*.

168 Id., *Nota inedita*, 1958, quot., (my translation).

> Such an inquiry coincides with the effort of a *dialectical philosophy*.
>
> A *dialectical philosophy* = cohesion of opposites, union through separation, reversal, truth in multiple articulations — hence, the revocation of an ontology of identity, and the idea of Being as that within which differences intertwine.
>
> Cf. *Dialectic*: Being admits the negative within itself.
>
> Cf. *Dialectic*: *Hyperconsciousness*: reflection is a return to the un-reflected – a separation that is also a union – the beginning is the end, and the end is the beginning.
>
> *Dialectic and subjectivity*: every dialectic is thus human, but not "human" in the sense of a separable possessor, *Man*. Cf. *Dialectic*: reason–non-reason.[169]

It is therefore necessary to understand, within this dialectical movement, that there is no origin, no end, nor, moreover, any proprietor who might govern the articulation of Being or who might perceive all its intrinsic possibilities and failures. Once again, then, it becomes necessary to confess, to *speak* the essential ambiguity of Being, insofar as "it (…) is of the same sort as the impartiality of past history, when it puts our former choices or the former doctrines beyond the true and the false for me the truth is this beyond the truth, this depth where there are still *several relationships* to be considered."[170]

Thus, the insistence on the partiality of truth, on its intrinsic ambiguity, fits perfectly with those "multiple entries of the world"[171] that give rise to a philosophy which, in order to gain access to these openings, must also become a non-philosophy – wonder, a beginning endlessly taken up again. All this implies that, just as there is never an absolute truth, there is neither a first nor a final word, nor – ultimately – a first question and a final answer; otherwise, Being would be deprived of its own dialectical movement, of its archaeological and teleological

169 Therein, quot., (my translation).
170 Id., *The Visible and the Invisible*, quot., p. 237.
171 Therein, p. 260.

dimension, of its very openness. For we must not forget – as Merleau-Ponty reminds us – that

> this frontal being before us – whether we posit it, whether it posits itself within us qua being-posited – is second by principle, is cut out upon a horizon which is not nothing, and which for its part is not by virtue of composition. One forgets that our openness, our fundamental relationship with Being, that which makes it impossible for us to feign to not be, could not be formed in the order of the being-posited, since it is this openness precisely that teaches us that the beings-posited, whether true or false, are not nothing, that, whatever be the experience, an experience is always contiguous upon an experience, that our perceptions, our judgments, our whole knowledge of the world can be changed, crossed out, Husserl says, but not nullified, that, under the doubt that strikes them appear other perceptions, other judgments more true, because we are within Being and because there is something.[172]

There is no final word capable of unveiling Being in its totality; nor is it possible to set a limit to philosophical discourse, for it – as interrogative thought – cannot cease to be "the most faithful witness of Being." Merleau-Ponty therefore invites us, at the close of his reflection – a reflection interrupted by his premature death – not to ignore the unsettling shadow of the unthought, to resist the temptation of letting thought rest upon itself, to maintain an unceasing vigilance.

This idea – or rather, this necessity of thinking – deeply marked those years in which the philosopher, summoned by the urgency of history, felt the duty to engage politically. In the background – that very background which, for Merleau-Ponty, makes the figure visible — there thus remains a philosophy understood as an exploration of the real and the possible, as interrogation of Being, as construction and deconstruction of sense and existence, and consequently as a continual decentering of the subject.

172 Therein, p. 127-128.

Reason then becomes a simple path traced within a web of sense and non-sense – a domain always already exposed to dissymmetry and conflict, because it has to do with the event, with heterogeneity, with otherness, with that which cannot be assimilated to itself. And precisely for this reason, this *ratio* does not exhaust the fecundity of experience, does not translate it into rigid logical-conceptual schemes expressed through absolutized linguistic structures that would claim to identify themselves with the very structures of reality; rather, it reveals itself as part of the world.

On these, which constitute the essential points of Merleau-Ponty's critique of traditional thought, the philosopher seems indeed to have traced a luminous path which — beginning with the phenomenology of genesis – has led us into a new and, in certain respects, uncertain and incomplete landscape of ontology.

The project of Merleau-Ponty, which this work has sought to bring to light, thus appears to be that of delineating the contours of a ""sensible reason" which, inhabiting the distance between Being and from Being, ultimately leads philosophy "to rediscover the [...] initial option which gave it birth, and to estimate the possibilities we have shut ourselves off from in becoming "Westerners" and perhaps reopen them."[173]

But if phenomenological-ontological inquiry legitimates itself as an attempt to unveil the mystery of the world, to grasp the sense of history and of the world in their nascent state, it cannot but continually call everything into question: *"the world,"* writes the philosopher, *"is always on its first day."*

This, then, implies that philosophy fulfills its task if – and only if – it experiences that wonder, that astonishment which is the continuous recovery of a beginning never surpassed, of a question never satisfied, and which, insofar as it becomes an answer, ceases to be what it ought to remain: interrogative thought. One must always begin anew, so that there may

173 Id., *Signs*, quot, p.139.

always be radical questioning, wonder, a beginning endlessly resumed.

It becomes clear, therefore, that Merleau-Ponty's thought never seeks – within any of his works – to establish the foundations of truth and certainty for our being-in-the-world; rather, it opens itself to that second, returning interrogation which must lie at the basis of our existence and, consequently, of our inscription within Being.

Hence, the deepest message of his ontology – the legacy he left us – is contained in this dense passage:

> The "originating" is not of one sole type, it is not all behind us; the restoration of the true past, of the pre-existence is not all of philosophy; the lived experience is not flat, without depth, without dimension, it is not an opaque stratum with which we would have to merge. The appeal to the originating goes in several directions: the originating breaks up, and philosophy must accompany this break-up, this non-coincidence, this differentiation. […] If the coincidence is never but partial, we must not define the truth by total or effective coincidence. And if we have the idea of the thing itself and of the past itself, there must be something in the factual order that answers to it. It is therefore necessary that the deflection *(écart),* without which the experience of the thing or of the past would fall to zero, be also an openness upon the thing itself, to the past itself, that it enter into their definition. What is given, then, is not the naked thing, the past itself such as it was in its own time, but rather the thing ready to be seen, pregnant— in principle as well as in fact— with all the visions one can have of it, the past such as it was one day *plus* an inexplicable alteration, a strange distance— bound in principle as well as in fact to a recalling that spans that distance but does not nullify it. What there is is not a coinciding by principle or a presumptive coinciding and a factual non-coinciding, a bad or abortive truth, but a privative non-coinciding, a coinciding from afar, a divergence, and something like a "good error".[174]

174 Id., *The Visible and the Invisible*, quot., pp.124-125.

Every philosophy that seeks to reabsorb the world of experience, to bring questioning to completion and thereby establish its own sovereignty over Being, remains at the level of superficiality and betrays its own vocation. For Merleau-Ponty, philosophy cannot be equal to this task until it becomes aware of its own "*enracinement*" in Being and is itself structured within this very act of awareness: "if philosophy wishes to be absolute, it contains itself."[175] It is "Being speaking within us," a perceptual faith questioning itself; it is a consciousness of that which precedes it, an attempt to let the silence of the world resound within its word, a bringing back of silence into this word. Far from asserting the superiority of Spirit, philosophy describes – on the contrary – the "passivity of our activity"[176]; as the poet Valéry wrote, a phrase repeatedly taken up in some of the essential moments of Merleau-Ponty's reflection, a "body of the Spirit"[177]. Spirit thus attempts to coincide with its carnal dimension, and this – rather than signifying a renunciation of its sovereignty – reveals, on the contrary, its most proper essence.

Philosophy therefore situates itself in the hollow of this chiasm, at this "*point de rebroussement*" where the mutual conversion of silence into speech and of speech into silence takes place; it seeks to grasp the equivalence between the being-sense of the world and the being-world of sense, between entering into oneself and going out of oneself; it seeks to fix the site where teleology becomes archaeology and vice versa.

But if this is indeed the case, ontology must install itself simultaneously on two sides: on the side of the world when it speaks, and, again, on the side of speech when it penetrates the thickness of the world. Indeed, it actively questions the very possibility of distinguishing two sides, a recto and verso

175 Therein, p. 167.
176 See Lisciani-Petrini, E., *Attività/Passività. L'invisibile di Merleau-Ponty*, quot., pp. 169-185.
177 Merleau-Ponty, M., *The Visible and the Invisible*, quot., p. 221.

of things, a subject and an object: it speaks of that exchange between us and the world – an exchange more profound than the very terms that intertwine within it.

When, in claiming to make itself a totality and to gather within the nets of sense the universe it names, philosophy attests to an originary relationship with itself and, consequently, with the world, an authentically philosophical word must be an interrogation of its own possibility – a search for what is "before" speech, yet without relegating this anteriority to the domain of non-sense: "the end of a philosophy," Merleau-Ponty wrote in *The Visible and the Invisible*, "is the accounting of its beginning."

In this final phrase, drawn from a working note dated February 1959, Merleau-Ponty brings into focus the entire meaning of his philosophical reflection. Here emerge, indeed, the most salient concepts of his path of thought: the circularity and reciprocal implication of the archaeological and teleological dimensions; the incompleteness of reduction as the rediscovery of vertical Being; and the ontological thematization of language, which is assigned the task of recounting always "the same tireless demand (…) from every lived thing (at times trifling), namely, the demand to be expressed."[178]

But what is "the accounting," precisely? When the French philosopher writes that "the end of a philosophy is the accounting of its beginning," he does not, of course, choose the term "*récit*" by chance. It possesses, as we shall attempt to show, a peculiar and decisive significance within the Merleau-Pontian discourse.

First of all, "*récit*" points back to an oral tradition – to the figure of Socrates, to the Plato of the *Dialogues* (grasping here, however, the symbolic meaning that Plato intended to give to these writings when he called them, precisely, "Dialogues"). Why does Merleau-Ponty entrust to narrative the highest philosophical task? Why does he not instead confer it upon

178 Id., *The prose of the world*, quot., p. 75.

writing, which is surely a more secure, more certain, more "true" mode of representation?

Here lies the fundamental point: writing, insofar as it is a representative expression, would in fact rise to the level of a pure *vision de survol* – an overview detached from the flesh of experience, a geometrical transcription codified once and for all, valid for every place and every time. It would once again be *mathesis universalis* – a mathematization of the world.[179]

The accounting, insofar as it is word, as mere voice, as λόγος προφορικός, stirred by the λόγος ενδιάθετος, and insofar as it has always already been exposed to its greatest danger – the forgetting of itself, the loss, the scattering of its own intentional threads that bind it inseparably to the world, the continual "sliding into the void," toward the "no longer having the place of a voice" of which Merleau-Ponty speaks in *The Visible and the Invisible* – expresses precisely the only possible mode of saying our original contact with Being.

From this it becomes clear in what sense the French philosopher entrusted to the "accounting" (*raconter*) the task of narrating the history of Being. And one also understands the ultimate aim of *The Prose of the World*, in which, to the "two-dimensional perspective" – that which "marks without ambiguity and without encroachment" all the "elements of the scenery" of the world – Merleau-Ponty opposes the "child's 'graphic narrations'". "The point – explains the philosopher – is that the aim is no longer to construct an 'objective' emblem of the spectacle [...] The aim is to leave on the paper a trace of our contact with this object and this spectacle". In other words, it is about "leaving a testimony and not any more of providing information". This "will arouse in us the profound schema which has settled us in our body and through the body in the world. It will bear the mark of our finitude, [...] it will lead us

179 Cf., in this regard, Sini C., *Filosofia e scrittura*, Laterza, Rome-Bari 1994.

to the secret substance of the object of which previously we possessed only the envelope".

> Two-dimensional perspective gave us the finitude of our perception, projected and flattened out, until it had become prose under the gaze of a god. The child's ways of expression, once they have been deliberately recaptured by an artist in a genuinely creative gesture, will, on the contrary, yield us the secret resonance through which our finitude opens up to the being of the world and becomes poetry.[180]

Paraphrasing Klee's words, the planimetric perspective represents the grafting of stability, the halting of movement, the fear of uncertainty; it thus escapes the mode of narration and clings to a space and time that are established, in themselves closed and resolved. Artistic language, by contrast, reveals the relativity of visible things, their inexhaustibility in relation to the totality of the visible, compelling one to proceed in fragments, and in the "remainder" of things "the intellect dies away miserably".[181] Therefore, back to Merleau-Ponty,

> The child's "graphic narrations" combine the successive scenes of a story into a single picture, drawing in the invariable elements of the scenery only once, or even drawing in just once each of the characters taken in an attitude suited to a given moment in the story. Thus the child is the sole hearer of the entire story at the time considered, and all the scenes communicate through the thickness of time, staking out the story further and further on. To the eye of the "reasonable" adult, who conceives time as a series of juxtaposed temporal points, such a narrative would be likely to seem broken and obscure. But if we follow time as we live it, the present still touches and still holds the past in its hand. The present is in a strange coexistence with the past, and only the ellipses in graphic narration can express

180 Cf. Merleau-Ponty, M., *The prose of the world*, quot., pp.150-151.

181 Cf. Klee, P., *Pedagogical Sketchbook*. London, Faber And Faber, 2000.

this movement of the story which encroaches its present onto its future in the way the "fold" expresses the coexistence of the invisible and visible aspects of the object or the secret presence of the object in the piece of furniture where it has been put away.[182]

182 M Merleau-Ponty, M., *The prose of the world*, quot., p. 151.

CHAPTER 2
THE DESCENT INTO THE FLESH OF THE WORLD

§ 1. *The Ontology of Flesh: Between Being and Being-of*

The gaze that grasps the advent of meaning by letting it come forth is not an act, but an adhering-to, a latent or operative intentionality within Being itself. Thus Merleau-Ponty can say, indifferently, that my seeing unfolds *from the midst of the visible, or from the midst of Being*[1]. I don't see Being *from the depths of nothingness, my vision of Being not forming itself from elsewhere; Being no longer being before me, but surrounding me and in a sense traversing me.*[2]

One must not conceive, says Cavaleri, our relation to Being as a relation between inside and outside, between fullness and emptiness taken in any absolute sense. I do not perform any opening toward this Being; rather, I *am of it* and *am within it*. My relation to the world – or to Being – is itself openness (cf. *VI*, 59), and my life unfolds nowhere outside this 'strange domain' (cf. *VI*, 152 and 140). The openness to the world thus presupposes that the world is, and remains, an horizon – in the sense that the one who sees both belongs to it and dwells within it."[3]

---

1 Cf. Merleau-Ponty, M., *The Visible and the Invisible*, quot., pp.113-114.

2 Ibid.

3 Cavaleri, B. *Sguardi e parole: lo stupore. Il paesaggio della "nuova ontologia"*, "aut-aut", 1989, n. 232-233, p. 58. (my translation). See also: Calabrò D., *Maurice Merleau-Ponty e il "labirinto dell'ontologia"*, in "Chiasmi International", n. 1,

The introduction of the notion of *being-of* (*en-être*) therefore does not merely relegate that of *being-in-the-world* to the background. As a consequence of this shift, one observes not only a deepening and radicalization of all the preceding Merleau-Pontian categories, but also the emergence of new concepts. Indeed, the explicit thematization of the ontological discourse requires a complete restructuring of the very concepts and of the language itself, and – as Zamboni notes – "an explicit reformulation of the very idea of the subject: so that things are subjects, just as language is a subject, and as the embodied *I* is a subject. And in a certain sense, none of these truly is a subject, for they are all enveloped and enveloping within the flesh of the world."[4]

Merleau-Ponty's lexicon thus becomes enriched with new expressions – some as explicit unfoldings of earlier ones, others entirely new. The ontological elucidation that is absent from *Phenomenology of Perception*, and only foreshadowed in *Signs*, is taken up explicitly in *The Visible and the Invisible*. Naturally, this is a work still in progress, marked by its own incompleteness – a research through which we have sought to bring to light certain motives that seemed to us the most fertile, the most promising of further developments. Whereas perception had provided the axis around which the vast problematic of *Phenomenology of Perception* was articulated, in the final phase of Merleau-Ponty's thought it gives way to deeper motives: *the world is what I perceive, but as soon as we*

1999, pp. 153-167; Gambazzi P., *Fissione dell'essere, essenze e visibilità assoluta. Individuazione ed ecceità della cosa nell'ultimo Merleau-Ponty,* in "Chiasmi International", n. 1, 1999, pp. 253-273; Matsuba S., *Possibilité de la "communauté charnelle",* in "Chiasmi International", n. 11, 2009, pp. 401-417; Morris D., *The chirality of being: exploring a Merleau-Pontyan ontology of sense,* in "Chiasmi International", n. 12, 2010, pp. 165-182.

4 Zamboni, C., *Il linguaggio nella riflessione di Merleau-Ponty e i legami con lo strutturalismo*, "aut aut", 1989, n. 232-233, p. 42. (my translation).

examine and express its absolute proximity, it also becomes, inexplicably, irremediable distance[5].

To say that there is transcendence, points out Merleau-Ponty, *being at a distance, is to say that being (in the Sartrean sense) is thus inflated with non-being or with the possible, that it is not only what it is*[6]. Thus, to the enveloping weave of *being-in-the-world* there succeeds the assumption of a negativity that is not negation understood as an aporetic limit of existence, but rather as a radical and ineliminable divergence: "Everything comes to pass as though my power to reach the world and my power to entrench myself in phantasms only came one with the other; even more: as though the access to the world were but the other face of a withdrawal"[7].

The Being to which we have access does not, therefore, possess the tranquil and massive consistency of the Being of classical ontology – the guarantor of truth and certainty – but is rather the only kind of Being accessible to existential inquietude: a raw, wild, primordial Being, which is neither subject-being nor object-being, and which cannot be seen as pure plenitude: *"Our point of departure shall not be being is, nothingness is not nor even there is only being – which are formulas of a totalizing thought, a high-altitude thought – but: there is being, there is a world, there is something; in the strong sense in which the Greek speaks of*to legein, *there is cohesion, there is meaning"*[8]. Here is what, according to Merleau-Ponty, is not taken into account:

> That the presence of the world is precisely the presence of its flesh to my flesh, that I "am of the world" and that I am not it, this is what is no sooner said than forgotten: metaphysics remains coincidence. That there is this thickness of flesh between us and the "hard core" of Being, this does not figure in the definition:

---

5 Merleau-Ponty, M., *The Visible and the Invisible*, quot., p. 8.
6 Therein, p.181.
7 Therein, p. 8.
8 Therein, p.88.

> this thickness is ascribed to me, it is the sheath of non-being that the subjectivity always carries about itself. Infinite distance or absolute proximity, negation or identification: our relationship with Being is ignored in the same way in both cases. In both cases, one misses it because one thinks one will ensure it more effectively by approaching the essence or the thing as closely as possible. One forgets that this frontal being before us – whether we posit it, whether it posits itself within us qua being-posited – is second by principle, is cut out upon a horizon which is not nothing.[9]

"According to an expression from *Signs*," writes Lefort, "it is the Being glimpsed through the tremor of time; and, to use Heidegger's phrase, the Being of distances – that which can never be grasped, for it is, in every sense of the word, ahead of us: at once behind and before the point at which we stand, already uttered or already uttering."[10]

The preliminary, phenomenological questioning addresses Being, yet what it aims toward is, to borrow the words of *Signs*, the *raw Being*, incommensurable with respect to the representations that science constructs of it, that something that presents itself in every experience as an absolute of presence, the *Être vertical*, that which stands upright before my standing body – not the Being crushed and offered to the dreams of a sovereign consciousness, but the *Esprit sauvage*. Now, this dimension, this disclosure of the raw Being, can be accessed through a questioning devoid of presuppositions, through a faith, a *Urdoxa*, that comes before any reasoning. In short, to describe the enigma of the raw world, one cannot resort to reflection or to science; we cannot

> resort to any of those established "truths" which we count on each day, and which in reality teem with obscurities, [...] the temptation to construct perception out of the perceived [...] is

9 Therein, p.127.

10 Lefort, C. *L'idée d'être brut et d'esprit sauvage,* quot., p. 272 (my translation)

> quasi-irresistible. […] We will not admit a preconstituted world, a logic, except for having seen them arise from our experience of brute being, which is as it were the umbilical cord of our knowledge and the source of meaning for us.[11]

Hence, Merleau-Ponty excludes "perception" insofar as it presupposes a fragmentation of lived experience into discontinuous acts, or a reference to "things" whose status is left unspecified, or simply an opposition between the visible and the invisible. Not because these distinctions are meaningless, but because they would lead back into the very aporias from which one is attempting to escape.

Only "perceptual faith," on the other hand, allows us to transcend the antinomy of a philosophy that accounts for Being and truth but neglects the world, and thus uproots us from both Being and truth.

> If coincidence is lost – writes Merleau- Ponty – this is no accident; if Being is hidden, this is itself a characteristic of Being, and no disclosure will make us comprehend it.. […] If it is to be the immediate […] this means that there is no route from us to it and that it is inaccessible by principle. […] The moment my perception is to become pure perception, thing, Being, it is extinguished; the moment it lights up, already I am no longer the thing.[12]

The truth is that the experience of a coincidence can be, as Bergson says, only a partial coincidence – a coincidence that is always past or always future, an experience that recalls an impossible past, that emerges from Being or is about to merge into it, that belongs to it, yet is not Being itself, and therefore is not coincidence, not a real fusion, but rather a covering:

---

11 Merleau-Ponty, M., *The Visible and the Invisible*, quot., pp.156-157.

12 Therein, p.122.

> The "originating" is not of one sole type, it is not all behind us; the restoration of the true past, of the pre-existence is not all of philosophy; the lived experience is not flat, without depth, without dimension, it is not an opaque stratum with which we would have to merge. The appeal to the originating goes in several directions: the originating breaks up, and philosophy must accompany this break-up, this non-coincidence, this differentiation.[13]

It is here that Merleau-Ponty's ontology reveals all its originality. Being is not, in fact, that primordial absolute that underlies individual existents. According to Merleau-Ponty, there is no metaphysical difference between Being and the entities in which it manifests itself. Even if certain expressions might suggest otherwise – such as when he speaks of vertical Being or Being of transcendence – Merleau-Ponty's ontology excludes a metaphysical dualism within the domain of Being. Verticality and transcendence are to be understood, rather, as surpassing, as divergence, a perspectival depth, as a lacuna:

> This being – which is glimpsed through time's stirrings and always intended by our temporality, perception, and our carnal being, but to which there can be no question of our being transported because to abolish its distance would be to take away its consistency of being – this being "of distances" as Heidegger will put it, which is always offered to our transcendence, is the dialectical idea of being as defined in the Parmenides – beyond the empirical multiplicity of existent things and as a matter of principle intended through them, because separated from them it would be only lightning flash or darkness.[14]

To clarify the relation between Being and entities, Merleau-Ponty appeals to dialectical thought. Yet he speaks of a dialectic conceived not as an absolute law, but as a tendency, which he calls *hyperdialectic* to distinguish it from the flawed dialectic of reflective philosophy.

13 Therein, p. 147.
14 Id., *Signs,* quot., p.156.

Hyperdialetics, the philosopher specifies, is not to be understood as a presupposition for common skepticism or relativism, nor as the condition for surrendering to the realm of the ineffable. Rather, it is the awareness of the provisionality of all historical events, of their partiality and contingency. *Hyperdialectic* does not exempt us from the laws of time, yet it provides the consciousness that history is composed of moments that intersect and collide, never arriving at definitive syntheses: to do philosophy is to pose questions; a conclusive answer would kill philosophy. As the philosopher explains, metaphysics is not a construction of concepts by which we try to make our paradoxes less noticeable but is the experience we have of these paradoxes in all situations of personal and collective history; one cannot conceive of a response which would eliminate such an inquiry; one thinks only of resolute actions which carry it further. Metaphysics is not a knowledge come to complete the edifice of knowledges but is lucid familiarity with whatever threatens these fields of knowledge and the acute awareness of their worth.[15] Metaphysics thus gives way, observes Nepi, to a semantic ontology in which signs never refer to a transcendent reality, constituting instead – on the model of linguistics – a diacritical system that can be endlessly decomposed and recomposed"[16].

Merleau-Ponty's perspective thus emerges in its essential features. The general framing of an interrogative thought becomes evident from the cited passage; the object of the questioning is contingency, that contingency which finds its organ in bodily perception, of which the philosopher provided a masterful phenomenology, and which, in *The Visible and the Invisible*, extends to the ontological plane.

The ontological trajectory traced by Merleau-Ponty should not, in our view, be understood as a negation of his previous

---

15 Id., *Sense and Non-sense,* quot., pp.95-96.

16 Nepi, P., *Merleau-Ponty tra il visibile e l'invisibile,* quot., p. 46 (my translation).

phenomenological path; on the contrary, it seems to us its logical continuation and necessary development. Indeed, the very demand for a deeper radicalization of the concept of *être-au-monde*, which had emphasized man's existential inherence in the world, drives the philosopher toward the search for a more primordial domain that allows one to ascend to that being of implication and latency – a horizon permanently present, yet also – as we have seen – a divergence and radical distance.

Thus arises a thought free from the categories of subject and object, a thought of indivision in contact with the variety of beings, a thought of latency devoted to approaching without ever fully reaching, illuminating without ever entirely bringing out of shadow, drawing on the source of perception, on the primordial relation it establishes with the world – a general experience of Being that does not rely on the distinction between positive and negative, between *in-itself* and *for-itself*. This, then, is the core of Merleau-Ponty's discourse as it unfolds through the passage from the phenomenology of genesis to the ontology of the "*être brut*" or "*sauvage*"[17].

From this emerges, then, that enigmatic world which must first be naïvely frequented[18] and only afterward critically interrogated – a world where the visible intertwines with the invisible. Thus, Merleau-Ponty's ontological elaboration manifests primarily as a questioning of our spontaneous vision

17 See Barbaras R., *Merleau-Ponty auu limites de la phénoménologie*, in "Chiasmi International", n. 1, 1999, pp 199-211; Stawarska B., *Dialogue at the limit of phenomenology,* in "Chiasmi International", n. 11, 2009, pp. 145-156.

18 *Naive frequenting of the world* is taken from Merleau-Ponty's posthumous volume, The Visible and the Invisible. "To frequent" is a verb of supple semantic depth: it indicates a constancy, the repetition of a presence within the mobility of an alternation between continuity and novelty, between initiative and abandonment. It is a kind of frequenting that does not obey a plan, that has in wonder the surrender to the primordial; an ingenuous engagement with the world, and thus a participation – beyond any critical exercise – in a vital and lived context in which we find ourselves involved.

of the world, of our pre-categorial perceptual experience. This is an important factor of continuity with his earlier phenomenologically inspired position, which presupposed the conviction that philosophy is rooted in lived experience. In *The Visible and the Invisible*, Merleau-Ponty reiterates that philosophy is above all the critical explication of this original experience, prior to thematic knowledge – that is, the *Urdoxa*: this naïve and unshakable "perceptual faith" in the world, which interrogates and returns upon itself. Within the continuity of this idea of philosophy, perception too is reaffirmed as that which "continues to (…) be experienced anew in each of our perceptions," the "primordial We", "the archetype of the originating encounter."[19] This is why "perception as an encounter with natural things is at the foreground of our research, not as a simple sensorial function that would explain the others but as the archetype of the originating encounter, imitated and renewed in the encounter with the past, the imaginary, the idea[20].

As C. Lefort suggests, "If there is need of a recommencement, it is therefore in a wholly new sense. It is not a matter of clearing out ruins in order to lay a new foundation; it is rather a matter of recognizing that, whatever we may say about being, we inhabit it with our whole selves, our labor of expression is still an installation in it, finally our interrogation is, for the same reason, without origin and without termination, since our questions always arise from older questions and since no answer can dissipate the mystery of our relation with being. […] the philosopher who frees himself from the myth of the 'root' resolutely accepts being situated in this midst and having to start from this 'some point or other.'"[21]

19 Merleau-Ponty, *Signs,* quot., p. 175, and *The Visible and the Invisible*, quot., pp. 51 and 158.

20 Id., *The Visible and the Invisible*, quot., p. 158.

21 Lefort, C., *Foreword* to Merleau-Ponty, M., *The Visible and the Invisible*, quot., p. XXVI.

Thus, philosophy is a questioning directed toward the functioning, pre-categorial life, in the trust that access to the original is possible. Yet access to the original does not entail a mystical fusion with the immediate – a thesis that the philosopher himself never endorsed, consistently refusing to follow this path in certain Bergsonian literature. This is a decisive point that must be clearly emphasized, in order to avoid attributing to Merleau-Ponty's thought movements that are not its own – as Invitto does when he writes: "if a criticism can be made of the late Merleau-Ponty, it is not that of idealism, but that of a kind of mysticism.""[22] Indeed, in our view, Invitto's thesis does not hold, for it is Merleau-Ponty himself who keeps well clear of mystical thought; not by chance, in a famous passage from *In Praise of Philosophy and Other Essays*, and precisely referring to Bergson, he writes:

> This secondary, laborious, rediscovered naïveté does not merge us with a previous reality, does not identify us with the thing itself, without any point of view, without symbol, without perspective.[23]

And moreover:

> What we propose here, and oppose to the search for the essence, is not the return to the immediate, the coincidence, the effective fusion with the existent, the search for an original integrity, for a secret lost and to be rediscovered, which would nullify our questions and even reprehend our language. […] Philosophy […] is a return upon itself and upon all things but not a return to an immediate— which recedes in the measure that philosophy wishes to approach it and fuse into it. The immediate is at the horizon and must be thought as such; it is only by remaining at a distance that it remains itself.[24]

---

22 Invitto, G. *Merleau-Ponty e la filosofia come vigilanza,* Milella, Lecce 1982, p. 212 (my translation).

23 Merleau-Ponty, M., *In Praise of Philosophy and Other Essays,* quot., p. 18.

24 Cf., Id., *The Visible and the Invisible*, quot., pp. 121-123.

From all this, it becomes clear how the new interpretative key to the original, elaborated in Merleau-Ponty's later works, is the concept of *flesh*, the true "sesame" – as Tilliette has said[25] – of Merleau-Ponty's ontology.

In Merleau-Ponty's sense, *flesh* designates the shared, deep layer of human and worldly sensitivity; it indicates that they are drawn from the same "stuff." It is the atmosphere, not perceptible as a tangible presence, that envelops and makes all realities possible, since they are precisely perspectives, forms, figures, symbols carved out of the flesh of the world. It is the element – in the Presocratic sense of the term – in which we are immersed, in that it constitutes the possibility of encountering anything else, without being itself an experienceable thing; it is, in short, the latency and possibility of every presence. This does not mean that it is the foundation of the totality of entities, but it operates within them as a power of productivity.

In the "cosmology" outlined in *The Visible and the Invisible*, the entities of the world partake of the polymorphic flesh as of the possibility that animates their infinite differentiation. This process of individuation arises anew from the perennial outburst of Being, from a dehiscence of the flesh, culminating in a doubling whereby the sensible becomes sentient and a chiasmatic relation is established between them.

Participants in the same flesh, the body of man and that of the world are not two separate realities but intertwined and enveloping one another: I unite with the things of the world by becoming world and making them flesh, in a chiasmatic relation that is not one of precedence or subordination, but of mutual transgression and surpassing. This means that the original relation between the bodily *I* and the world cannot be resolved either into a dualistic opposition or into a mere abstract identity but must be conceived as a thickness of flesh that ensures that "the presence of the world is precisely the

25 Tilliette, X., *Merleau-Ponty ou la mesure de l'homme,* Edition Seghers, Paris 1970, p.138.

presence of its flesh to my flesh"[26]. This invisible yet ever-present fabric is the ultimate gain of an inquiry that has become phenomenological-ontological.

With *flesh*, Merleau-Ponty is convinced that he has reached a dimension hitherto unexplored, long hidden within the tradition of Western thought. His endeavor, then, is to describe the carnal texture of the world without resorting to the classical categories of philosophical knowledge.

The flesh of the world – this raw, unobjectifiable Being – escapes the dualism of matter and spirit, nor can it be understood as substance. The only concept from the philosophical tradition that comes close to it is, as has already been noted, the Greek notion of *element*, "in the sense it was used to speak of water, air, earth, and fire, that is, in the sense of a *general thing,* midway between the spatio-temporal individual and the idea, a sort of incarnate principle that brings a style of being wherever there is a fragment of being. The flesh is in this sense an 'element' of Being"[27].

As an element, flesh is thus best understood—according to a metaphor much loved by Merleau-Ponty – as an *atmosphere* that radiates everywhere and is therefore simultaneously individuality and universality.

§ 2. *The permanent chiasm*

The notion of *being-of* and of *flesh* has already helped us to outline the meaning of the relation between Being and existents. Since it is meant to replace the category of being-in-the-world, the concept of *being-of* must therefore account for the new relation that, once the ontological discourse has been made explicit, comes to be established between consciousness

26 Merleau-Ponty, M., *The Visible and the Invisible*, quot., p.127.
27 Therein, p.139.

and the world. As we have seen, the category of *being-of* thus constitutes the radicalization of this relation.

A relation that, in *The Visible and the Invisible*, definitively takes on the figure of the *chiasm*. The I–world relation, as it appears in this work, confirms above all that the explicit shift toward the ontological perspective does not entail, for Merleau-Ponty, a complete abandonment of the earlier categories. The theme of the transcendental reduction, for example, is taken up once again and expressed with greater attention to the existential aspects of the relation between consciousness and Being. "One will say, then, that before the reflection, and in order to make it possible, a naive frequenting of the world is necessary"[28]. The phenomenological formula "reduction to the life-world" takes on here less technical, yet undoubtedly more evocative accents. One speaks of an "ingenuous familiarization with the world," where the intentionality of the phenomenological reduction is preserved in its fundamental claims, yet where the relation between the I and the world undergoes precisely that radicalization which the ontological discourse demands.

Merleau-Ponty seems to have grasped the risk of maintaining the reduction at the merely formal level of method, with an undue hypostatization of the transcendental consciousness; and it is for this reason that, in the new perspective, the inherence of the I and the world is emphasized with particular clarity. The I and the world are now affirmed in their inseparable reciprocity.

The phenomenological and ontological perspectives merge into a unified and original semantic nexus. Here, according to Rigobello, "the Husserlian lesson meets that of Heidegger. In the background, not too distant, remains the charm of the subtle Bergsonian influence, which the contribution of

28 Merleau-Ponty, M., *The Visible and the Invisible*, quot., p.51.

German phenomenology rearticulates within a different epistemological framework."[29]

The being of the I and the being of the world – whose reciprocity and inherence the notion of *being-of* deepens – are both elucidated by the concept of flesh, understood as the proper dimension of Being, its metaphysically originary element. Flesh must therefore be conceived in light of the Being that incarnates itself – *in the sense of visibility and concreteness* – in every existent. And it is precisely insofar as they are the in-carnation of the originary Being that the I and the world belong to one another, that they are intertwined in a chiasmatic involvement. The role of the body as mediation between the I and the world is likewise clarified and deepened through the category of flesh. "The thickness of the body – says the philosopher – far from rivaling that of the world, is on the contrary the sole means I have to go unto the heart of the things, by making myself a world and by making them flesh."[30] It is, in fact, precisely the flesh of my own body that enables me to understand and to know the flesh of the world. In light of this deepening, the phenomenological description becomes more intricate: the body is both seeing and visible, sensing and sensible. My right hand touching my left – perceiving it as an object at the very moment in which the latter, in turn, feels itself touching, that is, as a perceiving subject – is the event in which the intertwining and the chiasm of being are expressed in their most exemplary form. And this occurs without the body once again becoming the locus in which a constituting subject is absolutized, since the sentient–sensible chiasm no longer grounds, as before, the originary dimension of the body; that dimension is now given by the flesh of the world as the primordial weave, the texture, lining, and fabric of every entity. Both the I and the world participate in the

29 Rigobello, A. *L'impegno ontologico,* Armando, Rome 1977, p. 109 (my translation).

30 Merleau-Ponty, M., *The Visible and the Invisible*, quot., p.135.

flesh; they are of it: it is the metaphysical element that allows every existent to relate to Being, to *be-of-it*.

Thus, the reality of the body as a chiasm of sentient and sensible, seeing and visible, provides the interpretive key to the chiasm in which the most profound meaning of Merleau-Ponty's ontology is condensed – the chiasm of the visible and the invisible.

In this respect, it should be noted that whereas previously the intertwining had been primarily between the sentient and the sensible and therefore situated on the level of perception, it is now chiefly between the seeing and the visible. What does this shift from perception to vision mean? According to Nepi, "in contrast with perception, which tends to close itself within the narrow confines of the perceiving subject and the perceived object – ultimately finding itself faced with the very opposition between subject and object that one intended to overcome – vision seems instead to presuppose an immediacy of relation between the seer and the visible. This is perhaps the reason why, at the moment when the ontological problem is explicitly thematized, Merleau-Ponty passes from perception to vision".[31] On this point, however, we believe that Sordini's explanation[32] brings out more clearly the transition from seeing sensibility to sensing visibility, and from perceptual vision to ontological vision, by insisting on the essential connection – identified by Merleau-Ponty – between "eye" and "spirit." "For they are joined," Sordini writes, "by a natural pact. The distinction is thus only apparent: thought is always incarnate. The act of seeing wrenches the spirit away from its tautological relation to itself and initiates it into the mystery of alterity: 'the eye performs the prodigy of opening for the soul what the

31 Nepi, P., *Merleau-Ponty tra il visibile e l'invisibile,* quot., p. 46 (my translation).

32 Cf. Sordini, A. *Pittura metafisica nell'ultimo Merleau-Ponty,* in *Il corpo vissuto*, Il Saggiatore, Milan 1979, pp. 195-201 and p. 156.

soul itself is not: the bright realm of things – and their god, the sun.'"[33]

Vision is, then, the dimension of existence in which the manifestation of Being is most immediately realized. And this is because vision takes place within an atmosphere in which human subjectivity becomes maximally rarefied – within that limit-experience where anthropology already borders on ontology. Indeed, vision, as the chiasm of seer and seen, is to be understood as an interpretive model of Being itself, and not merely of the human. "It suffices for us for the moment to note that he who sees cannot possess the visible unless he is possessed by it, unless he *is of it,* unless, by principle, according to what is required by the articulation of the look with the things, he is one of the visibles, capable, by a singular reversal, of seeing them – he who is one of them"[34].

Thus, by understanding vision as the event that allows us to witness from within the fission of Being, Merleau-Ponty arrives at that metamorphosis of things into things-seen which opens a breach into the heart of Being precisely insofar as it shows more than the things themselves: it renders present what is.

The dimension that vision unveils – the visible – is therefore not space *in itself*, continuous and homogeneous, positive-objective, a Cartesian space without hiding places, where every point is identical to itself and occupies a fixed position determined by the three-dimensional axes of length, width, and depth. The Cartesian idealization of space conceals its primordial depth: in this depth things spill over into one another, overlapping reciprocally, participants in the dimensionality of carnal being – that primordial depth in which everything is simultaneously present and from which the three dimensions are extracted as distinct systems of measure.

33 Therein, pp. 200-201 (my translation).

34 Merleau-Ponty, M., *The Visible and the Invisible*, quot., pp. 134-135.

But then, on closer view, the chiasm of seer and seen presupposes an even more radical intertwining: that between the visible and the invisible. Merleau-Ponty is keenly aware that, once situated at the level of the originary, we cannot stop halfway. It is the very world of the visible that opens us onto the world of the invisible. "Between the alleged colors and visibles, we would find anew the tissue that lines them, sustains them, nourishes them, and which for its part is not a thing, but a possibility, a latency, and a *flesh* of things."[35]

Within the visible there is always a natural displacement: the emergence of negativity, or depth. In the very fabric of the visible there is need for this differentiation, this hollow that will be vision. If Being did not harbor this weakness – if it were a Being without constraint – it would be identical to an absolute and undifferentiated nothingness in which nothing could ever take place. The divergence between the visible and the invisible (like the divergence between background and figure) is the "site" where meaning comes to be.[36]

Precisely this, according to Merleau-Ponty, is the heart – or the centrality – of Being. One must think the center, the heart of Being, as the point at which Being intersects with nothingness, the point of reversal, the fold in which the visible, turning back upon itself, comes to be seen and makes me a seer. *Central* therefore means the *central point of the chiasm*, the point of divergence between Being and its negativity,[37]

---

35 Therein, pp. 132-133.

36 In these terms, according to Carbone, "Merleau-Ponty's philosophy takes shape as a philosophy of *divergence*, as a white space in which an unspoken – otherwise inexpressible – finds indirect expression. In Merleau-Ponty's conception this *divergence* is what is traced within the chiasm of the visible and the invisible." (The quotation appears in: Carbone, M., *Il sensibile e il desiderio. Merleau-Ponty, Lyotard e la pittura*, "aut-aut," 1989, n. 232–233, p. 84.)

37 Merleau-Ponty takes up the theme of negation from Hegel, affirming that "All the greatl philosophical ideas of the past century [...] had their beginnings in Hegel. [...] And again, referring to

between Being and its individuation. But then the invisible is what makes vision possible.

In everything I see and everything I say there is always included what I do not see and what I do not say: what lies

---

Hegel, he writes: An absence of being would have to come into the world, a nothingness from which being would be visible, such that consciousness of life, taken radically, is consciousness of death. Even the doctrines which would imprison us in our racial or local peculiarities and hide our humanity from us can only do so-since they are doctrines and propaganda-by forsaking the immediate life and borrowing shamefully from consciousness of death. […] There are two ways of thinking about death: one pathetic and complacent, which butts against our end and seeks nothing in it but the means of exacerbating violence; the other dry and resolute, which integrates death into itself and turns it into a sharper awareness of life. […] To plumb our awareness of death, we must transmute it into life, "interiorize" it, as Hegel said. […] There is no being without nothingness, but nothingness can exist only in the hollow of being, and so consciousness of death carries with it the means for going beyond it. […] Domination, sadism, and violence destroy themselves just as consciousness of death goes beyond itself. If each participant in the duel of consciousnesses, of fraternal enemies, succeeded in fatally wounding the other, nothing would be left; there would not even be a place for that hatred of the other and that affirmation of self which is the principle behind the struggle. Thus, the man with the most exact awareness of the human situation is not the master (since the master pretends ignorance of the foundation of being and communication underlying the play of his despair and pride) but the slave. The slave has been truly afraid, has given up trying to conquer by the sword, and he is the only one with experience of death because he alone has known the love of life. The master wants to exist for no one but himself, but in fact he seeks recognition of his mastery from someone and so is weak in his strength. The slave consents to exist only for others, but nevertheless it is he who chooses to go on living on these terms, and he therefore has strength in his weakness. Since he is better acquainted with man's vital core than the master, it is he who will finally have the only possible mastery […] His life is more frankly rooted in the world than is the master's, which is precisely wb,y he knows better than the master what death means". The two quotations are taken from Merleau-Ponty, M., *Sense and Non-sense*, quot., pp. 63 and 66-69.

between things and what lies *between* words. The visible and language reveal, by their very presence, a secret absence, a latency—something from which alone visibility and meaning can emerge. All this allows us, finally, to discern in Merleau-Ponty's late meditation a concept of difference as *divergence*, which takes on a distinctive significance: it expresses an originary dimension, not of substance, but of latency—the perennial source of its inexhaustible possibilities of intramundane meaning.

The fundamental thought of ontology is thus now configured not as the thought of being as foundation, but as that of its fragmentation, of the interminable differentiation instituted by the outburst of Being, which is the ultimate intranscendable event to which philosophical questioning of the meaning of the world arrives. But where, within this teleology of being, does the sentient–sensible chiasm in which the dehiscence of flesh culminates find its place? This chiasm marks the beginning of the process in which carnal being occurs, becomes consciousness, and takes on *logos*. The *logos* that unfolds in this chiasm is, consequently, a wild *logos*, insofar as it does not originate from an act of the rational Self, nor does it designate a supernatural light preexisting the world; rather, it is the perceptual sense that arises with the dehiscence of flesh, a natural light. *Logos*, therefore, is Being itself returning upon itself, making itself present and, so to speak, illuminating itself, making itself visible. Yet is the Being of latency, the invisible fold of being, merely what is currently unseen due to purely empirical situations, or is it that which can never, under any condition, be seen? ome pages from the preface to *Signs*, written in the same years in which the philosopher composed the manuscript of *The Visible and the Invisible*, offer some guidance in answering. In them, one can indeed read: "To see is as a matter of principle to see farther than one sees, to reach a latent existence. The invisible is the outline and the depth of the visible. The visible does not admit of pure positivity any more than the invisible does."[38] And, with an obvious polemical reference to Sartre, he continues: "It would be better to speak of "the visible and the invisible," pointing out that they are not contradictory, than to speak of "being and nothingness." One says invisible as one says immobile – not in reference to something foreign to movement, but to something which stays still. [...] The philosophy which lays bare this chiasma of the visible is the exact opposite of a philosophy of God-like survey. It plunges into the perceptible, into time and history, toward their articulations. It does not surpass them through forces it has in its own right; it surpasses them only in their meaning."[39]

For Merleau-Ponty, overview thinking is any cognitive activity that is content with grasping only the surface of reality, through conceptual abstractions and logical categories whose epistemological rigor is inversely proportional to their existential content. The dialectic between the visible and the invisible can also be understood in this context. The invisible is then construed as spirit opposed to body, idea opposed to matter.

Yet, according to Merleau-Ponty, there is a more authentic way to read the relation between the visible and the invisible. The invisible – excluding the banal solution that it is merely what is currently unseen but could at any moment enter the sphere of visibility – is, as we have said before, the latent dimension of the visible, its structural framework, its hidden being. In the concept of the invisible, therefore, converge in a single theoretical nexus the ontological categories of transcendence, latency, absence, and dehiscence, upon which the ontology of carnal being is founded. This ontology, by not thematizing being as fullness and identity but conceiving it as *Abgrund*, is wholly differentiated (which was precisely Merleau-Ponty's aim) from the onto-theological conception

38 Merleau-Ponty, M., *Signs,* quot., pp.20-21.
39 Therein, p.21.

of being, which instead posits being as the guarantee and foundation of reality, ultimately serving to justify the existing order and becoming functional to the strategy of preserving and securing the sense that "we are at home."[40] Here, in this impossibility of arrival, of feeling chez-soi, lies the validity and richness of Merleau-Ponty's philosophical inquiry.

§ 3. *The envelopment between eye and mind, body and world*

"We see things; we agree about them; we are anchored in them; and it is with "nature" as our base that we construct our sciences. Cézanne wanted to paint this primordial world and his pictures therefore seem to show nature pure"[41]. […] "In search of depth Cézanne seeks this deflagration of Being, and it is all in the modes of space, in form as much as anything."[42]

These two judgments on Cézanne mark the beginning and the end of a research trajectory that engaged with pictorial expression, under the persistent conviction that the hesitant murmur of colors agitating within the painting is not merely the ornament of the depicted design but reveals the pulsation of the originary – it is the manifestation of the "aconceptual universality" of wild *logos*.

During the phenomenological phase of the 1940s, painting occupied a prominent place among Merleau-Ponty's areas of investigation, which he conceived as the expressive operation in which the secret of the "strange adherence" of the tacit cogito to the lived world manifests itself. In the later period, when the phenomenology of perception gives way to the new perspective of the indirect ontology of flesh, painting becomes central to the French philosopher's interest, as he recognizes in it the privileged experience illuminating access to Being.

---

40 Id., *The Visible and the Invisible*, quot., p. 104.

41 Merleau-Ponty, M., *Sense and Non-sense*, quot., p. 13.

42 Id., *Eye and Mind,* in *The Primacy of Perception,* quot., p.180.

What we now propose, then, is to focus on the specificity and, at the same time, the lines of continuity in the ontological interpretation of painting developed in *L'œil et l'esprit (Eye and Mind)*, in relation to the previous phenomenologically inspired interpretation, taking it as a testing ground for the heuristic potential of Merleau-Ponty's final theoretical program.

Indeed, even in his pre-ontological writings, the philosopher turns to painting to make it a philosophical experience, insofar as it is recognized as the most expressive form of tacit language: in the inquiry into the connection between language and lived body, painting appears as the mute form of artistic expression that most faithfully expresses the originary perceptual experience, as the perpetual recreation of the lived world unfolding from the subjectivity of the body itself. In the body, as the opening and vehicle of being-in-the-world, the secret of the expressive operation and the peculiar style of painting is thus discerned, which does nothing other than amplify and articulate its expressive powers, enhancing its visual perception: if the painter renders visible what would otherwise remain occluded from consciousness, his creative power derives from the capacity of the lived body to transcend itself toward the world and toward other subjects.

Only because pictorial expression is rooted in the bodily dimension of gesture and perception does it have the ability to disclose the world of life, allowing us to witness from within the genesis of the raw sense that occurs in it. The "early" Merleau-Ponty centers his attention especially on Cézanne, because he recognizes in his style the manifestation of pictorial capacity as an openness to the world-of-life and the possibility of making its functional sense visible: his work is the path of return to precategorial, mute, and solitary experience, upon which art and the entire order of culture are founded. Cézanne paints the primordial world in its nascent state, creating while constantly assuming the attitude of the first human painting, because the Nature he faces is always "on the first day"; for this reason, in his eyes, his work is never complete. In short, he opens himself to the unsurpassable fullness of the world-of-life, and his vision "penetrates right to the root of things beneath the imposed order of humanity."[43]

The interpretation of painting articulated in *Eye and Mind*, while reaffirming the idea of art as questioning and as a continuous rebirth, is nonetheless reshaped, on the whole, by the two factors that mark the final and difficult years of Merleau-Ponty's inquiry: the opening toward the demands of structuralism, and the turn toward an ontological perspective.

The influence of the first of these motifs has been identified and examined by Lefeuvre.[44]. He rightly observes that, if in *Le doute de Cézanne* and *in Le langage indirect et les voix du silence* Merleau-Ponty had shown how pictorial expression was the prolongation of the bodily gesture and thereby formed one with the expressive miracle of the lived body – thus arriving at a grounding of the order of culture in the world of perception – he had nevertheless not adequately identified what is specific to art, failing to clarify in what respect it truly differs from the perceptual sphere. The encounter with Saussure's semiology now provides him with the means to fill this gap, allowing him to conceive painting as an autonomous system of differentiated

43 Id., *Sense and Non-sense*, quot., p.16. About the Merleau-Ponty's thought on Cézanne, see in particular: Duportail G.F., *La sainte victoire de Cézanne,* in "Chiasmi International", n. 17, 2015, pp. 225-236; Kakuni T., *Le corps aux limites de la rapprésentation. Théorie du corps et de la peinture chez Merleau-Ponty,* in "Chiasmi International", n. 12, 2010, pp. 203-214; Casey E.S., *Visibilizing the invisible in painting,* in "Chiasmi International", n. 19, 2018, pp. 239-253; Mercury J.-Y., *L'expressivité chez Merleau-Ponty. Du corps à la peinture*, L' Harmattan, Paris 2000; Carbone M., *Ai confini dell'esprimibile. Merleau-Ponty a partire da Cézanne e da Proust*, Guerini e Associati, Milano 1990; Id., *Mondo estetico, arte, pensiero*, Guerini e Associati, Milano 1996; M. Carbone e D. Levin, *La carne e la voce. In dialogo tra estetica ed etica*, Mimesis, Milano 2003.

44 Lefeuvre, M., *Merleau-Ponty au-delà de la phénoménologie du corps, de l'être et du langage,* Université de Lille III, 1977.

signs. This does not mean, however, that Merleau-Ponty severs the bond between art and corporeity. In *Eye and Mind*, the thesis is, in fact, reaffirmed that painting rests upon the body, that the enigma of vision – the spontaneous concordance between the subjective dimension of seeing and the objective presence of the thing seen—has its hidden ground in the body's capacity to furnish, within its own flesh, an interior equivalent to the sensory features of the visible world.

> The painter "takes his body with him," says Valéry. Indeed, we cannot imagine how a *mind* could paint. It is by lending his body to the world that the artist changes the world into paintings. To understand these transubstantiations, we must go back to the working, actual body—not the body as a chunk of space or a bundle of functions but that body which is an intertwining of vision and movement.[45]

"The creative power of painting," Mancini explains, "lies in the pictorial signs themselves and is no longer reducible merely to the capacity of the bodily self to transcend toward the world and to disclose its immanent meaning. But it is above all in its intertwining with the mediation on Being that the interpretation of painting achieves a qualitative leap. [...] The approach to painting presented in Eye and Mind is grounded in a new phenomenological survey of the pre-predicative experience of seeing, interpreted in the light of the notion of flesh, which constitutes the core of Merleau-Ponty's ontology. "[46]

Flesh, in fact, is also the pivot around which the complex analysis of vision and the dialectic of visible and invisible is organized. It is constitutive both of the visibility of the thing and of the perceiving body itself, for it is their shared element

45 Merleau-Ponty, M., *Eye and Mind,* in *The Primacy of Perception,* quot., p.162.

46 Mancini, S. *Il narcisismo della carne. Merleau-Ponty e l'interpretazione ontologica della pittura,* "aut-aut", 1984, n. 203-204, pp. 143-144 (my translation).

and thus allows a relation to be established between the seer and the visible. In *The Visible and the Invisible*, it designates the common layer of the sensitivity of the self and of the events of the world, the very fabric from which they are taken as differentiated realities. It is not perceivable as a tangible presence, for it is that which envelops and renders all realities co-possible within the same horizon of depth.

The thickness of flesh, in short, is what gives the visible that density which allows it to be seen by a body, and at the same time it is what arrests the gaze of that body. In the structure of vision, flesh manifests itself as the enveloping of the visible over the body of the seer: it is, therefore, what makes their contact possible, and it ensures that in their reciprocal exchange an anonymous Visibility is formed, which properly belongs neither to the perceiving body nor to the visible things.

By positing anonymous visibility as the expression of flesh, the event of vision comes into focus: visible and seer reveal themselves as two aspects of a single phenomenon. Indeed, just as the body itself is simultaneously sensitive and sentient, and thereby functions as a universal measure, so too – inasmuch as it is a seeing body – it is on the one hand a visible among others, yet on the other hand it is that visible which sees itself in conjunction: being at once seer and visible, it functions as an archetypal visible. Vision, the central cavity of the visible, is then nothing other than this invisible doubling of the visible in the seer, and it comes to configure the very notion of flesh.

Understanding vision in the light of flesh, Merleau-Ponty conceives it as an event that exceeds the strictly anthropological horizon: vision, this "magical" doubling of the visible in the seer, is understood as a paradox of Being, and not solely of man, because "we do not make (vision) for it makes itself in us"[47]. But above all, by positing vision as the primordial mode of bodily being and grounding it in the paradox of the seeing

47 Merleau-Ponty, M., *Eye and Mind,* in *The Primacy of Perception,* quot., p.167.

and visible body, Merleau-Ponty avoids conceiving it as an operation of thought, as a mere "overview" of the world.

To vision, understood as the return of the visible upon itself, the philosopher ascribes an insurmountable narcissism, due to the fact that the seeing body desires to see itself as seeing, and in this it is seduced by itself. This narcissism does not manifest solely at the level of the self, but reveals the original narcissism of carnality itself, which, in its process of differentiation, seeks to reach itself, to see itself, and thus to make itself visible. The narcissism of vision is therefore rooted in the reversibility of raw carnality: this is why the vision exercised by the seer is simultaneously imposed upon the seer by the things themselves.

This, moreover, is attested by many painters who have indeed acknowledged having experienced the sensation of being looked at by the things in the act of painting them, of migrating into them: therefore, painters "have so often liked to draw themselves in the act of painting (they still do – witness Matisse's drawings), adding to what they saw then, what things saw of them. It is as if they were claiming that there is a total or absolute vision, outside of which there is nothing and which closes itself over them."[48] Within this interpretive context, marked by symbolic reference, lie the brief notes on the mirror, whose power to magically transform things into spectacles and spectacles into things, me into the other and the other into me, is recognized. This power rests upon the primordial "reflectivity" of carnate sensibility, which it translates and doubles: "Artists have often mused upon mirrors because beneath this 'mechanical trick,' they recognized, just as they did in the case of the trick of perspective, the metamorphosis of seeing and seen which defines both our flesh and the painter's vocation."[49]

48 Therein, p.169.

49 Therein, pp.168-169. Reference to the "trick of perspective" can be found in Delaunay, R., *Du cubisme* à l'art *abstrait,* S.E.V.P.E.N.,

From all these themes, it is clear that original visibility belongs

> properly neither to the body qua fact nor to the world qua fact—as upon two mirrors facing one another where two indefinite series of images set in one another arise which belong really to neither of the two surfaces, since each is only the rejoinder of the other, and which therefore form a couple, a couple more real than either of them.[50]

But all of this, then, raises again the problem of the invisible, since it is only against the backdrop of the invisible that the specular interplay of visibility can occur. Once more, it becomes evident that "Merleau-Ponty's endeavor, as Mancini observes, is to interpret the invisible without falling into a new dualism; it is therefore a question of identifying, through a new phenomenological description sensitive to its ontological implications and stripped of any idealistic legacy, the invisible nuclei of meaning – which are nothing other than the pivots of carnate being – without opposing this dimension of negativity and depth to that of visibility, in the preliminary recognition that this 'void' of the invisible is firmly lodged within visible being. Moreover, he seeks to make thought emerge within the infrastructure of vision, avoiding both the endorsement of an empiricist genesis of thought and subordination to the dualistic opposition of thought and extension inaugurated by Descartes. Thought, indeed, is understood as the realization of an invisible that constitutes the power of original visibility, insofar as it is the fulfillment of the *inaugural there-is*, and this fulfillment occurs through a process of sublimation of flesh."[51]

Paris, 1957.

50 Id., *The Visible and the Invisible*, quot., p.139.

51 Mancini, S., *L'ontologia indiretta dell'ultimo Merleau-Ponty*, in VV.AA., *Merleau-Ponty: Esistenza, filosofia, politica*, ed. by Invitto, G. Guida, Naples 1982, p. 73 (my translation). See also at this point: Larison M., *L'Être en forme. Dialectique et phénoménologie dans la derneère philosophie de Merleau-Ponty*,

Moreover, as Robinet writes, "by granting thought the privilege of seeing at the expense of vision, Descartes turned his back on the visible with the inevitable and necessary aim of freeing the spirit from empiricism in order to advance science. [...] Ambiguity, ubiquity, and unreality of vision are nothing but the fallacious consequences of an order of life for which the intellect is not made, a mere decree of an omnipotent and hidden will."[52]

It thus becomes clear – the divergence hinted at from the outset – that separates *Eye and Mind* from Merleau-Ponty's earlier writings on painting. Painting is no longer understood as the subjective reprise, within the order of sense, of the lived world; rather, it is called upon to contribute to disclosing our access to raw being and to its wild logos. Its ultimate significance does not consist in being a human representation of the world of life, but in presenting itself as the place where being exceeds presence, manifesting itself as visible being: "The painter's vision is not a view upon the outside, a merely 'physical-optical' relation with the world. The world no longer stands before him through representation; rather, it is the painter to whom the things of the world give birth by a sort of concentration or coming-to-itself of the visible"[53].

Depth and movement, color and line, are thus situated within an ontological interpretive framework, as centers of radiance of carnate being, and the painter and his world reveal themselves as two differentiations, placed in a reversible chiasmatic relation, participants in the teleology of "natural light" and its mute historicity.

Editions Mimesis, Milan 2016.

52 Robinet, A., *Che cosa ha veramente detto Merleau-Ponty*, it. transl. Ubaldini Editore, Rome 1973, p. 43 (my translation)

53 Merleau-Ponty, M., *Eye and Mind,* in *The Primacy of Perception,* quot., p.181. Reference to the physical-optical' relation to the world is to Paul Klee.

CHAPTER 3
"A WORLD FILLED WITH MEANINGS"

§ 1. *Spoken Cogito and Tacit Cogito*

Merleau-Ponty's reflection on a carnal ontology, by bringing to light the ambiguity inherent in the relation between the visible and the invisible, ultimately unfolds into a "semantic ontology"[1]: there is, in fact,

> a *being* of language in which the enigma of being is repeated, because beyond the movement of the pure significations there remains the silent mass of the discourse, that which is not of the order of the sayable, and because the greatest merit of expression is to disclose this continuous passage from the word to being and from being to the word, or this double openness of the one upon the other. To think through this exchange is no doubt what *The Visible and the Invisible* was to devote itself to [...] there is no frontier between language and the world.[2]

But in what sense can one speak of a "semantic ontology"? In the sense that, in *The Visible and the Invisible*, language becomes a mode of being in accordance with the ontology of the flesh and of the chiasm. If the idea (the sense) is a "taking up from the visible," if Being differentiates itself and

---

1 The expression "semantic ontology" appears in P. Nepi, *Merleau-Ponty tra il visibile e l'invisibile*, quot.; see in particular pp. 27–33 and 132–144. See at this point: Vanzago L., *Leggere Il visibile e l'invisibile di Merleau-Ponty*, Ibis, Como-Pavia 2020.

2 Merleau-Ponty, M., *The Visible and the Invisible*, quot., pp. 153-154.

"exposes" itself by becoming meaning, gesture, figure, then language is

> the totality of what is said, the integral of all the differentiations of the verbal chain; it is given with the words for those who have ears to hear. And conversely the whole landscape is overrun with words as with an invasion, it is henceforth but a variant of speech before our eyes, and to speak of its "style" is in our view to form a metaphor. In a sense the whole of philosophy, as Husserl says, consists in restoring a power to signify, a birth of meaning, or a wild meaning, an expression of experience by experience, which in particular clarifies the special domain of language. And in a sense, as Valery said, language is everything, since it is the voice of no one, since it is the very voice of the things, the waves, and the forests.[3]

Language, therefore, "is the voice of no one, since it is the very voice of the things"[4].

The world of language is, for Merleau-Ponty, the embodied world. The dialectic without synthesis between myself and things, structured by the chiasm, finds its parallel in the dialectic without synthesis between myself – who, in responding to the appeal of things, operates unwittingly within language – and the speaking language, the "*langage langagier*", which comes to materialize itself in the direction of sense.

Thus, on this side of conventional means of expression – which manifest my thought to the other insofar as they are already given in me as in him, since for each sign there are meanings – one must recognize, according to Merleau-Ponty, a primordial operation of signification, one that already takes place at the level of the life-world, thanks to transcendental intersubjectivity. Against every philosophy of deduction and construction, Merleau-Ponty therefore once again privileges the antepredicative, as an appropriation of the sense of the world, opened stratigraphically toward successive explication.

3 Therein, p.155.
4 Ibid.

> The spoken word is a genuine gesture, and it contains its meaning in the same way as the gesture contains its. This is what makes communication possible [...]. We live in a world where speech is an *institution*. For all these many commonplace utterances, we possess within ourselves ready-made meanings. They arouse in us only second order thoughts; these in turn are translated into other words which demand from us no real effort of expression and will demand from our hearers no effort of comprehension. Thus, language and the understanding of language apparently raise no problems. The linguistic and intersubjective world no longer surprises us, we no longer distinguish it from the world itself, and it is within a world already spoken and speaking that we think. We become unaware of the contingent element in expression and communication, whether it be in the child learning to speak, or in the writer saying and thinking something for the first time, in short, in all who transform a certain kind of silence into speech. It is, however, quite clear that constituted speech, as it operates in daily life, assumes that the decisive step of expression has been taken. Our view of man will remain superficial so long as we fail to go back to that origin, so long as we fail to find, beneath the chatter of words, the primordial silence, and as long as we do not describe the action which breaks this silence.[5]

But this "semantic ontology" is only the final outcome of a reflective trajectory on language that proceeds in parallel with the itinerary already outlined in the previous chapters. The treatment of language, therefore, does not receive in Merleau-Ponty a single, univocal solution; rather, it acquires different valences depending on the various stages of his path of thought. For the sake of clarity, we could schematically define it as gestural in the *Phenomenology of Perception*, *diacritical* in *Signs* and in *The Prose of the World*, and finally ontological in *The Visible and the Invisible*. The basic framework is certainly provided by the *Phenomenology of Perception*, even though, in the later works, Merleau-Ponty's discourse deepens, becoming far richer and far more problematic. It is

5 Id., *Phenomenology of Perception*, quot., pp. 213-214.

therefore from that fundamental early work that it is once again appropriate to begin. "Dealing with language," Scotti writes – referring precisely to the *Phenomenology of Perception* – "Merleau-Ponty no longer recognizes it as a phenomenon of extrinsic explication by means of a system of conventional signs, but as a consciousness expressing itself through the body, without this implying, even implicitly, any form of dualism. On the contrary, with the description of the act of signification, he seeks to overcome definitively the classical dichotomy of subject and object. The word is not a simple social instrument but expresses the moment in which the I no longer adheres immediately to its environment but transcends nature and the world of objects."[6]

Merleau-Ponty thus begins his phenomenological analysis of language by seeking to recover the "speaking subject" against the classical, empiricist and intellectualist conceptions. According to empiricism, in fact, language is reduced to the mere existence of verbal images, that is, to traces left within us by the words we have uttered or heard. Language would thus be marked by the absence of a "speaking subject"; it would be a third-person process, on a par with other organic processes: there is no one who speaks, but only a flux of words that come forth without being governed by the intention to speak. Yet the analysis of aphasia, which concerns the very intentionality of language and shows how it becomes compromised through illness, reveals the untenability of such a thesis. However, not even the intellectualist theory, according to Merleau-Ponty, resolves the problem of language, for the solution cannot be found by passing from thesis to antithesis. Whereas for empiricism what is essential is the reproduction, the externalization of the word, for intellectualism the word is merely the outer casing of true denomination and of authentic speech, which is a wholly interior operation, one that unfolds

6 Scotti, G. *Originarietà e relazione in Merleau-Ponty*, "aut-aut" 1957, n° 39, p. 306 (my translation).

exclusively at the level of intellectual representation. And yet the two conceptions agree on this: for both, the word does not possess meaning in itself:

> In the first case this is obvious since the word is not summoned up through the medium of any concept, and since the given stimuli or 'states of mind' call it up in accordance with the laws of neurological mechanics or those of association, and that thus the word is not the bearer of its own meaning, has no inner power, and is merely a psychic, physiological or even physical phenomenon set alongside others, and thrown up by the working of an objective causality. It is just the same when we duplicate denomination with a categorical operation. The word is still bereft of any effectiveness of its own, this time because it is only the external sign of an internal recognition, which could take place without it, and to which it makes no contribution.[7]

The word is without meaning insofar as it does not possess meaning; it is thought that has meaning, and the word remains an empty shell. It is only an articular, sonorous phenomenon, or the consciousness of this phenomenon. In any case, language is nothing but an external accessory of thought. In the first conception we stand on this side of the word as meaningful; in the second, beyond it. In the first, there is no one who speaks; in the second, there is indeed a subject, but it is not the speaking subject – it is the thinking subject. Arguing in particular against intellectualism, which reduces language to an exterior expression of thought, Merleau-Ponty also writes:

> If speech presupposed thought, if talking were primarily a matter of meeting the object through a cognitive intention or through a representation, we could not understand why thought tends towards expression as towards its completion, why the most familiar thing appears indeterminate as long as we have not recalled its name, why the thinking subject himself is in a kind of ignorance of his thoughts so long as he has not formulated them for himself, or even spoken and written them, as is shown by the

7 Merleau-Ponty, M., *Phenomenology of Perception*, quot., p. 205.

> example of so many writers who begin a book without knowing exactly what they are going to put into it. A thought limited to existing for itself, independently of the constraints of speech and communication, would no sooner appear than it would sink into the unconscious, which means that it would not exist even for itself.[8]

Therefore, far from being a simple sign of objects and meanings, speech inhabits things and is the vehicle of meaning: "speech, in the speaker, does not translate ready-made thought, but accomplishes it." […] There is thus, […] a *thought in speech* the existence of which is unsuspected by intellectualism"[9]; meaning is captured in the word that is its external existence: the word is the worldly structure of thought. So, "the word and speech must somehow cease to be a way of designating things or thoughts and become the presence of that thought in the phenomenal world, and, moreover, not its clothing but its token or its body."[10] Thought, therefore, is not anything interior; it does not exist outside the world and outside the word.

Thought and expression thus constitute themselves simultaneously, just as our body lends itself immediately to a new gesture in the acquisition of habit.

The word is therefore a bearer of meaning, exactly as the body is in itself a "nascent symbolism." For this very reason Merleau-Ponty, following Saussure, distinguishes between *langue* and *parole*, between the crystallized and sedimented language-system and the event-word, which is a modulation of our coexistence and co-belonging to the world. Consequently, Merleau-Ponty criticizes the interpretation of the word as a "conventional sign" and, in the *Phenomenology of Perception*, moves instead toward an interpretation of the word as a "gestural sign":

8 Therein, p. 206.
9 Therein, pp. 207 and 209.
10 Therein, p. 211.

> It is through my body that I understand other people, just as it is through my body that I perceive 'things'. The meaning of a gesture thus 'understood' is not behind it, it is intermingled with the structure of the world outlined by the gesture, and which I take up on my own account. It is arrayed all over the gesture itself. […] The linguistic gesture, like all the rest, delineates its own meaning. […] It seems at first impossible – says Merleau-Ponty – to concede to either words or gestures an immanent meaning, because the gesture is limited to showing a certain relationship between man and the perceptible world, because this world is presented to the spectator by natural perception, and because in this way the intentional object is offered to the spectator at the same time as the gesture itself. Verbal 'gesticulation', on the other hand, aims at a mental setting which is not given to everybody, and which it is its task to communicate. But here what nature does not provide, cultural background does. Available meanings, in other words former acts of expression, establish between speaking subjects a common world, to which the words being actually uttered in their novelty refer as does the gesture to the perceptible world. And the meaning of speech is nothing other than the way in which it handles this linguistic world or in which it plays modulations on the keyboard of acquired meanings. I seize it in an undivided act which is as short as a cry […] One can see – continues Merleau-Ponty – what there is in common between the gesture and its meaning, for example in the case of emotional expression and the emotions themselves […] Strictly speaking, therefore, there are no conventional signs, standing as the simple notation of a though: pure and clear in itself, there are only words into which the history of a whole language is compressed, and which effect communication with no absolute guarantee, dogged as they are by incredible linguistic hazards.[11]

"The specificity of language," Bonomi observes, "must be preserved, because emotion itself does not depend causally on neural processes, but is rather a certain way of structuring them. If, on the one hand, conventional signs no longer exist, on the other, natural signs do not exist either, for they would

11 Therein, pp. 216-218.

presuppose an organization that makes definite gestures correspond to certain layers of consciousness, whereas emotion is a form and not an epiphenomenon. This is why, Bonomi concludes, in human behavior one cannot isolate a component that depends simply on nature and another that depends on the cultural world."[12]

Insisting on the gestural or existential meaning of speech, Merleau-Ponty therefore emphasizes how it primarily expresses the subject's position in the world:

> The phonetic 'gesture' brings about, both for the speaking subject and for his hearers, a certain structural co-ordination of experience, a certain modulation of existence, exactly as a pattern of my bodily behavior endows the objects around me […] The meaning of the gesture – outlines Merleau-Ponty – is not contained in it like some physical or physiological phenomenon. The meaning of the word is not contained in the word as a sound. […] If it is authentic speech, just as gesture endows the object for the first time with human significance, if it is an initiating gesture. Moreover, significances now acquired must necessarily have been new once. We must therefore recognize as an ultimate fact this open and indefinite power of giving significance— that is, both of apprehending and conveying a meaning—by which man transcends himself towards a new form of behaviour, or towards other people, or towards his own thought, through his body and his speech.[13]

Similarly, phenomenological analysis shows that behind language there is not a transcendent thought, but that thought transcends itself in the word. And the word is transcendence because spoken language never expresses the totality of experience but is always overdetermined with respect to it; because there is a fundamental obscurity in what is expressed, and thought is never absolutely self-transparent. Language

---

12 Bonomi, A. *Esistenza e struttura. Saggio su Merleau-Ponty*, quot., p. 78 (my translation).

13 Merleau-Ponty, M., *Phenomenology of Perception*, quot., pp. 225-226.

surpasses us, not only because the use of words always presupposes a great number of thoughts that are not actual and that each word summarizes, but also because these very thoughts were never pure, insofar as in them meaning already exceeded the signifier.

This relationship between thought and language is called, in phenomenological terms, *Fundierung*. That is, the founding term – be it time, the un-reflected, fact, language, or perception – is first in the sense that the founded is given as a determination or an explication of the founding term, without ever being able to reabsorb it. The founding term, however, is not first in the empiricist sense, and the founded is not its mere derivative; rather, the founding term manifests precisely through the founded.

Our situation is thus the following: in order to know that we are thinking, we must first actually think: "the aim at a transcendent objective and the view of myself aiming at it, the awareness of the connected and of connecting are in a circular relationship"[14]. The question, however, is to understand how subjectivity can be at once dependent and yet inalienable. Merleau-Ponty seeks to clarify this problem through language, which allows us to grasp, beyond mere expression, a truth separable from it and of which expression is only the contingent manifestation. Given this function, language does not transmit the finite perspective of my perception, but rather the sense that, in intention, exceeds that perspective. It is language, therefore, that fills perception, giving it meaning, because it says more than what it sees.

The dialectic of word and perspective is thus the very dialectic of infinity and finitude. And it is within this dialectic that the gap between certainty and truth, between the intention of truth and present certainty, between signifying and perceiving, comes into play – the disproportion between the word, which says Being, and the gaze, which remains

14 Therein, p. 466.

anchored to appearance and perspective. Within this gap, then, there emerge the residuals of non-sense, of the involuntary, of the irrational, whose attestation arises within the very horizon of consciousness itself, as part of its lived experience and of its radical ambiguity.

> The study of perception could only teach us a "bad ambiguity," a mixture of finitude and universality, of interiority and exteriority. But there is a "good ambiguity" in the phenomenon of expression, a spontaneity which accomplishes what appeared to be impossible.[15]

And then, in *The Visible and the Invisible*, one can read: *Perception has me as has language*[16]. But this power of language to express what exists, to open paths for thought, to disclose new dimensions, new landscapes, remains, despite everything, *obscure* and *invisible:*

> Behind the spoken *cogito*, the one which is converted into discourse and into essential truth, there lies a tacit *cogito*, myself experienced by myself. But this subjectivity, indeclinable, has upon itself and upon the world only a precarious hold. It does not constitute the world, it divines the world's presence round about it as a field not provided by itself; nor does it constitute the word, but speaks as we sing when we are happy; nor again the meaning of the word, which instantaneously emerges for it in its dealing with the world and other men living in it, being at the intersection of many lines of behaviour, and being, even once 'acquired', as precise and yet as indefinable as the significance of a gesture. The tacit *cogito*, the presence of oneself to oneself, being no less than existence, is anterior to any philosophy and knows itself only in those extreme situations in which it is under threat: for example, in the dread of death or of another's gaze upon me. […] The tacit *cogito* is a *cogito* only when it has found expression for itself.[17]

---

15 Id., *An Unpublished Text*, in *The Primacy of Perception*, quot., p. 11.

16 Id., *The Visible and the Invisible*, quot., p. 190.

17 Id., *Phenomenology of Perception*, quot., pp. 469-470.

Thus, while the spoken cogito is language, the tacit cogito is that un-reflected dimension which can become reflection only by taking the form of word: the movement by which the silent cogito comes to self-knowledge is the movement of linguistic signification. Indeed, Merleau-Ponty continues,

> if we were to make completely explicit the architectonics of the human body, its ontological framework, and how it sees itself and hears itself, we would see that the structure of its mute world is such that all the possibilities of language are already given in it. Already our existence as seers (that is, we said, as beings who turn the world back upon itself and who pass over to the other side, and who catch sight of one another, who see one another with eyes) and especially our existence as sonorous beings for others and for ourselves contain everything required for there to be speech from the one to the other, speech about the world.[18]

But then, just as had already occurred for other problematic orders, the reflection on the language of the body opens onto the ontological dimension. Indeed, all the Merleau-Pontyan discourse analyzed thus far ends up showing how there is a mutual interpenetration between the problem of Being and that of language. Hence, "The Saussurean analysis of the relations between signifiers and the relations from signifier to signified and between the significations (as differences between significations) confirms and rediscovers the idea of perception as a *divergence (écart)* by relation to a *level,* that is, the idea of the primordial Being"[19]; but it should also be noted that "The openness upon a natural and historical world is not an illusion and is not an a priori; it is our involvement in Being"[20], and what no reduction can reduce is the phenomenon of language, my inherent connection to a world *filled with meanings*.

From the ontological perspective clarified in *The Visible and the Invisible*, language establishes itself precisely as a

18 Id., *The Visible and the Invisible,* quot., p. 155.
19 Therein, p. 201.
20 Therein, p. 85.

pathway to the problem of Being, not because Being could be reduced to formal predication, but because language allows us to grasp and express its structures and to enter into contact with the intersubjective community – that is, to access a universe of raw being and coexistence. For this reason,

> Like the flesh of the visible, – writes Merleau-Ponty – speech is a total part of the significations, like it, speech is a relation to Being through a being, and, like it, it is narcissistic, eroticized, endowed with a natural magic that attracts the other significations into its web, as the body feels the world in feeling itself. In reality, there is much more than a parallel or an analogy here, there is solidarity and intertwining: if speech, which is but a region of the intelligible world, can be also its refuge, this is because speech prolongs into the invisible, extends unto the semantic operations, the belongingness of the body to being and the corporeal relevance of every being, which for me is once and for all attested by the visible, and whose idea each intellectual evidence reflects a little further.[21]

Rather than possessing the secret of being-in-the-world, language is itself a world; it is itself a being-in-the-world and a being raised to the second power, because it does not speak into the void but speaks of Being and of the world. The mistake of the various philosophies of language lies in enclosing language, as if it spoke only of itself. In reality, lived experience is a spoken-experience, and language is not a mask of Being but – provided that one grasps it in its radicality – the most faithful witness of Being. Consequently, for Merleau-Ponty, language, if one considers it as already constituted, is only the secondary and empirical operation of translation, of codification and decodification; it is artificial language, the technical relation between a sound and a meaning that are joined solely by convention.

If, instead, one considers speaking speech, the enfolding of the visible and the lived onto language, and of language

21 Therein, p. 118.

onto the visible and the lived; the exchanges between the articulations of its mute landscape and those of its utterance; finally that operative language which has no need to be translated into significations and thoughts, that language-thing which brings to the surface all the deep relations of the lived through which it was formed – then, in that case, language is a universal problem, the ultimate theme of philosophy.

Merleau-Ponty, therefore, once again transcends the "world of consciousness" and its acquired, constructed significations, placing the problem in its radical terms and opening onto an original perspective – onto a Being that stands behind verified being, and that is that raw spirit out of whose horizon constructed objects and significations emerge. The problem is thus brought back to its ontological matrix, to the dimensionality of Being and to our relationship with it[22].

§ 2. *Constituted Speech and Silence*

> The philosopher speaks, but this is a weakness in him, and an inexplicable weakness: he should keep silent, coincide in silence, and rejoin in Being a philosophy that is there ready-made. But yet everything comes to pass as though he wished to put into words a certain silence he hearkens to within himself. His entire "work" is this absurd effort. He wrote in order to state his contact with Being; he did not state it, and could not state it, since it is silence. Then he recommences…. One has to believe, then, that language is not simply the contrary of the truth, of coincidence, that there is or could be a language of coincidence, a manner of making the things themselves speak — and this is what he seeks. It would be a language of which he would not be the organizer, words he would not assemble, that would combine through him by virtue of a natural intertwining of their meaning, through the occult trading of the metaphor – where what counts is no longer the manifest meaning of each word and of each image, but the

22 See the interesting monographies on Merleau-Ponty and latest ontology: Taddio, L., *Maurice Merleau-Ponty,* Feltrinelli, Milano 2024 and Vanzago L., *Merleau-Ponty*, Carocci, Roma 2012.

> lateral relations, the kinships that are implicated in their transfers and their exchanges. It is indeed a language of this sort that Bergson himself required for the philosopher. But we have to recognize the consequence: if language is not necessarily deceptive, truth is not coincidence, nor mute.[23]

This passage, one of Merleau-Ponty's most celebrated, condenses with evocative accents the final outcome of his thought, in which all the themes gradually taken up by the philosopher converge in an exemplary synthesis..

Sini, a keen interpreter of Merleau-Ponty, precisely referring to this passage, comments: "Language and truth constitute the hallmark of philosophy; which strives to capture the muted words that Being murmurs. Merleau-Ponty's project was '*to return to the very origin of the history of speech or, rather, to tear speech out of history,*' tracing back to the great silence of the world in and from whose flesh the word bursts forth: Our understanding of man will remain superficial until we trace back to this origin, until we rediscover, beneath the murmur of words, the primordial silence, until we describe the gesture that breaks this silence.

"For this reason," Sini continues, "Merleau-Ponty sought to '*consider speech before it has been pronounced, against the ground of the silence which precedes it, which never ceases to accompany it, and without which it would say nothing. Moreover, we should be sensitive to the thread of silence from which the tissue of speech is woven.*'"[24]

It is in silence and from silence that I, the world, and language emerge in their unitary correlation: enigmatic poles of the same linguistic or linguistic-sign game, masks of the Same, approximated distances in which, as Peirce maintained, each is where it functions – that is, where it announces itself

---

23 Merleau-Ponty, M., *The Visible and the Invisible,* quot., p. 125.

24 Cf., Sini, C. *Il silenzio e la parola*, quot., pp. 10-13 (my translation). Quotes are taken from M. Merleau-Ponty, *The Prose of the World,* quot., pp. 5 and 46.

and resonates. The philosopher's word, that word which, as Merleau-Ponty says, must "accompany" this *contingent* game, is thus an echo of the world: always distanced from its very point of eruption "Being no longer being *before me,* but surrounding me and in a sense traversing me […], my vision of Being not forming itself from elsewhere, but from the midst of Being"[25]. Thus, speech speaks only from the silence of the world, that silence which it preserves and carries with it. It cannot fully say it (for it speaks and resonates in an echo without origin, or whose origin is precisely the silence to be spoken, the silence of speaking); yet it always says it, each time recommencing the rebound of the echo that originates the world and marks the distance."[26]

In *The Visible and the Invisible*, the great silence of the world is explicitly taken as the incomparable setting that serves both as the background and as the crucible for the first miraculous eruption of the human voice and of the word: a sign that is woven into the flesh of the world and yet surpasses it, envelops it, involves it. This revelation of carnal correspondence is always, at the same time, the manifestation of distance and of difference – of an absence and an unbridgeable remoteness, of elusiveness and of a grasp maintained at a remove. Language is thus an oriented and in-carnated system, one that arises from the "silent world" of the body, from the "need," from the "passion" for speaking.[27]. And here is the articulation of Merleau-Ponty's reflection in its various junctures:

> Language lives only from silence; everything we cast to the others has germinated in this great mute land which we never leave […] the factual presence of other bodies could not produce thought or the idea if its seed were not in my own body […] a sublimation of the flesh […] will be mind or thought. […]. Yet

25 Merleau-Ponty, M., *The Visible and the Invisible*, quot., p.114.
26 Sini, C. *Il silenzio e la parola*, quot., pp. 10-13 (my translation).
27 Cf. Merleau-Ponty, M., *Signs*, quot., p. 17 and *The Visible and the Invisible*, quot., p. 126.

> this flesh that one sees and touches […] will pass definitively beyond the circle of the visible, […] (this) going beyond (goes towards) […] a negativity that *comes to the world.* […] The invisible dimension is hidden within the visible; it is what gives it depth and weight, just as there is a mutual covering and enfolding between silence and language. […] Language arises from… and speaks… all the silences; it articulates what the world, Being, the sensible, and silence have always meant to convey […] The word introduces a ferment of transformation into silent vision; it brings a disruption into it. Just as flesh (or the gaze) emerged from the intertwining of the visible and the invisible and was nothing but their remainder, their *between*, so the word is born at the point where language and silence intersect. The second term is the depth, the thickness of the first; it appears only within it and continues to envelop it. Silence is the zero-point of language, not its opposite […]. Silence is not absolute, just as language is not absolute: the former teems with words—there are voices within silence—while the latter is woven with silences. One never manages to say everything, to exhaust silence; this is why the effort of expression is always on its first day, beginning anew with every human being and never coming to an end. Silence seeks words, and the word seeks silence. […] philosophy is language and nonetheless consists in rediscovering silence […]. What will this silence be? […] This reversal itself – *circulus vitiosus deus*- is not hesitation, bad faith and bad dialectic, but return to åigh' the abyss.[28]

The continuous cycle of rejection and retrieval of silence signifies that there are two silences between which the word is situated – diacritically – whereby, according to Lefort, "it gives expression to an experience that is mute and ignorant of its own meaning, but only in order to make that experience appear in its purity; it does not break our contact with the things, but it draws us all things only in order to awaken us to the truth of their presence and to render palpable their relief and the tie that binds us to them. At least such is the speech that speaks in conformity with its essence and, where

28 Merleau-Ponty, M., *The Visible and the Invisible*, quot., pp. 126, 145, 176, 179.

philosophical discourse is concerned, that does not cede to the vertigo of eloquence, does not wish to suffice to itself or close in upon itself and upon its sense, but opens upon and leads to the outside. But if speech, which is born from silence, can seek its conclusion in silence and make that silence not be its contrary, this is because between experience and language there is, in principle, exchange; it is because experience is not something one could coincide with, because it bears a transcendence, since already, in itself, it is differentiation, articulation, structuration, and because in some way it calls for language; it is because language is also experience, because there is, [....] a *being* of language in which the enigma of being is repeated, because beyond the movement of the pure significations there remains the silent mass of the discourse, that which is not of the order of the sayable, and because the greatest merit of expression is to disclose this continuous passage from the word to being and from being to the word, or this double openness of the one upon the other"[29].

In this sense, one may speak of a reversibility of language: thought and word mutually anticipate one another, continuously substituting for one another.

> Operative language makes us think, and living thought magically finds its words [...], [....], Expressive operations take place between thinking language and speaking thought; not, as we thoughtlessly say, between thought and language.[30]

Thus, "speaking becomes both signifying and founding, in the sense that it can no longer be a simple statement, a *Satz*; it is necessary, Merleau-Ponty says, that it be a thinking word, without reference to a *Sachverhalt* – speaking and not merely language."[31]. This because, precisely, "speaking that

29 Lefort, C. *Foreword* to Merleau-Ponty, M., *The Visible and the Invisible*, pp. XXVIII-XXIX.

30 Merleau-Ponty, M., *Signs*, quot., p. 18.

31 Costantino, S. *La struttura del linguaggio in Merleau-Ponty,*

constitutes, *in front of* myself as a signification and a subject of signification, a milieu of communication, an intersubjective diacritical system which is the spoken tongue *[la langue*] in the present, not a "human" universe, an objective spirit)"[32]. As we can read in *The Visible and the Invisible*,

> Being opens itself to me in a *logos*, and the 'natural magic' of the word opens me to this *logos* of Being. The operative essence is behind the operative speech, [...] a common nervure of the signifying and the signified. [...] Speech prolongs into the invisible, extends unto the semantic operations the belongingness of the body to being and the corporeal relevance of every being. [...] As the world is behind my body, the operative essence is behind the operative speech also, the speech that possesses the signification less than it is possessed by it, that does not speak *of it,* but speaks *it,* or speaks *according to it,* or lets it speak and be spoken within me, breaks through my present.[33]

The word, now intertwined with the flesh, thus appears in an explicit ontological dimension: the *chair langagière*. In this sense, instead of revealing to us the secret of the being of the world, language is itself a world, it is itself a being – a world and a being raised to the second power, because "it does not speak in a vacuum, since it speaks *of* being and *of* the world and therefore redoubles their enigma instead of dissipating it"[34]. We are thus led back to the ontological perspective that takes language as the pathway of access to the problematic of Being.

> It is the error of the semantic philosophies to close up language as if it spoke only of itself: language lives only from silence; [...] But, because he has experienced within himself the need to speak, the birth of speech as bubbling up at the bottom of his mute experience, the philosopher knows better than any-

---

Corso Editore, Ferrara 1976, p. 94 (my translation).

32 Merleau-Ponty, M., *The Visible and the Invisible*, quot., p. 175.

33 Therein, p.118.

34 Therein, p.96.

> one that what is lived is lived-spoken, that, bom at this depth, language is not a mask over Being, but [...] the most valuable witness to Being, that it does not interrupt an immediation that would be perfect without it, that the vision itself, the thought itself, are [...] structured as a language, are *articulation* before the letter, apparition of something where there was nothing or something else.[35]

However, "one would be mistaken in thinking that the articulatory structure of the vision of thought is an end in itself [...]; it is instead necessary to specify that this structure possesses a value of reversibility, at least with regard to speaking speech, between the visible and the lived upon language, and language upon the visible and the lived; and since this reversibility extends to the articulations of the 'mute landscape' and of the word, since true language is the operative one – 'which needs to be translated into signification and thoughts' insofar as it is already itself signification and thought, in the sense that it presents itself *d'emblée* as language-thing – then the *logos* of language is an absolutely universal theme; it is (as has already been said) the theme of philosophy."[36]

> Philosophy itself is language, rests on language; but this does not disqualify it from speaking of language, nor from speaking of the pre-language and of the mute world which doubles them: on the contrary, philosophy is an operative language, that language that can be known only from within, through its exercise, is open upon the things, called forth by the voices of silence, and continues an effort of articulation which is the Being of every being.[37]

This is because, as Lefort observes, "I engage with Being before knowing it; my questions emerge prior to becoming

35 Therein, p.126. Reference to vision and thought "structured as a language" is taken from Jacques Lacan.

36 Costantino, S. *La struttura del linguaggio in Merleau-Ponty*, quot., pp. 98-99 (my translation).

37 Merleau-Ponty, M., *The Visible and the Invisible*, quot., pp.126-127.

philosophical interrogation. Thus, I must bear in mind what exists before any thesis, before any reflection; I must maintain this engagement with Being amid the semi-silence and semi-darkness of consciousness."[38] All this implies, quite naturally, that the method of accessing Being is "indirect," "secondary"; "*One cannot make a direct ontology*"[39]

To utter the silent language of Being, to produce thereby, in a sort of detonation, the innate knowledge of the body through which Being manifests itself, to gut, so to speak, the mystery and essential opacity of things – this is the demand that characterizes the indirect method of the French philosopher. It is certainly visible, in the late Merleau-Ponty, a "strong Heideggerian temptation,"[40] insofar as philosophy is brought back to the "explication" and *"recollection" of Being:*

> The "amorphous" perceptual world that I spoke of in relation to painting – perpetual resources for the remaking of painting – which contains no mode of expression and which nonetheless calls them forth and requires all of them and which arouses again with each painter a new effort of expression – this perceptual world is at bottom Being in Heidegger's sense, which is more than all painting, than all speech, than every "attitude," and which [...] appears as containing everything that will ever be said [...]: it is the λόγος ενδιάθετος which calls for the λόγος προφορικός – [Iteration of the *Lebenswelt:* we are making a philosophy of the *Lebenswelt,* our construction (in the mode of "logic") makes us rediscover this world of silence.[41]

If it is through language that one speaks and says silence, so too is it through silence that one speaks and says that word which comes from absence. It is therefore, fundamentally, a matter of traversing silence: of experiencing it, of taking it

38 Lefort, C. *L'idée d'Être brut et d'Esprit sauvage,* in "Les Temps Modernes", n. 184-185, 1961, p. 271 (my translation).
39 Merleau-Ponty, M., *The Visible and the Invisible*, quot., p. 179.
40 cf. Derossi, G., *Dalla percezione alla visione. L'ontologia negativa dell'ultimo Merleau-Ponty*, "Filosofia", 1964, p. 346.
41 Merleau-Ponty, M., *The Visible and the Invisible*, quot., p.170

into account, and, in some way, of producing it. Silence is not mute word, but a word always already open to the slipping-away into nothingness. In this sense, paradoxically, "the dwelling of the human has truths richer than the saying of the word: silent enchantments that open the world long before it is reduced to that alternative of the 'public' and the 'private' that underlies all our forms of knowledge."[42] The dwelling place of man is, strictly speaking, always a utopian exodus that can never be reached because, as Derrida writes, "the look cannot abide."[43]

Then perhaps what a language without Voice might be—a word no longer founded upon any will-to-say – is what we must, today, learn to think with Merleau-Ponty and beyond Merleau-Ponty. This exigency echoes, for instance, in a scholar who has worked on language, though from a perspective not Merleau-Pontyan but Heideggerian: Giorgio Agamben, whose words we repeat here, writes: "To dwell in language without being summoned by any Voice, to simply die without being called by death, is perhaps the most abyssal experience; yet precisely for this reason, it is also, for humans, the most habitual experience, their ethos, their dwelling, which, in the history of metaphysics, always appears already demonically split into living and language, nature and culture, ethics and logic, and is therefore accessible only within the negative articulation of a Voice. And perhaps, only from the eclipse of the Voice, from the absence of a locus for language and the death of the Voice, does it become possible for humans to experience their own ethos in a way that is no longer merely sigetic."[44]

---

42 Sini, C. *Il silenzio e la parola*, quot., pp. 7-8 (my translation).

43 Derrida, J., *Speech and Phenomena, and Other Essays.* Northwestern University Press, Evanston, 1973, p. 104.

44 Agamben, G., *Il linguaggio e la morte*, quot., pp. 120-121 (my translation).

Would this invisible, this unsayable, this unheard-of be what always calls to emerge, between the white spaces of the page, in the late Merleau-Ponty?

The philosopher's premature death does not allow us to give a certain and definite answer, for we cannot know – nor even imagine – what the final outcomes of his thought would have been. Yet the questions with which his discourse comes to a close serve for us as *"signs,"* traces to be followed toward new – for now "invisible" – "full of meaning", events[45].

45 See at this point: Lanfredini R. and Vitale S., (eds.), *Pensare in ultimo. Merleau-Ponty dopo Merleau-Ponty*, Moretti & Vitali, 2014; R. Lanfredini, *Divenire di Merleau-Ponty. Filosofia di un soggetto incarnato*, Guerini e Associati, Milano 2011.

FINAL MEDITATIONS ON TIME, SPACE, AND MOVEMENT
Inspired by Unpublished Notes

The closing text of this volume unfolds around several unpublished manuscripts[1] by the French thinker. From his vast body of notes, the present inquiry focuses on those dated September 23 and October 6-7, 1958[2], as well as on the *Esquisse ontologique*, likely written in March 1959[3]. The

1 Before entering the heart of the analyses devoted to the *papiers inédits* under consideration, it is useful to offer a few brief indications regarding the overall organization of Merleau-Ponty's manuscripts. They have been divided into seven *boîtes* and catalogued according to a thematic rather than a strictly chronological order; that is, they follow the sequence of the works to which they belong and the progression of the Courses the philosopher taught at the Collège de France. Moreover, since October 6, 1995, thanks to the authorization granted by Mme Suzanne Merleau-Ponty, all the manuscripts have been made available for consultation in the Merleau-Ponty Collection of the Manuscripts Department of the Bibliothèque Nationale de Paris.

2 These are some of the notes that Claude Lefort chose not to append to the unfinished manuscript of *Le visible et l'invisible*, so as not to burden its structure. The notes dated 6 and 7 October 1958 have been placed in *Boîte* III, entitled "Le visible et l'invisible," under the heading "Notes inédites." The placement of the note from September 23, 1958, *Pour l'ontologie*, is as follows: *Boîte* IV, D.M.O., B.N.F. It forms part of the "Projet et plan et troisième ébauche" of the *Projets de livres* 1958–1960 (*Boîte* IV, D.M.O., B.N.F.).

3 The *Esquisse ontologique* constitutes the nucleus from which the broader manuscript entitled *Projets et plan du livre en vue depuis 1958* unfolds, a manuscript that itself belongs to the dossier "Notes pour choses à faire" (*Boîte* III, D.M.O., B.N.F.). It is the preliminary draft – an "outline", precisely – prepared with a view to the unfinished work *Origine de la vérité*, which later became *Le*

selection of these documents is by no means accidental; it is driven by a precise theoretical motivation: to show how, beginning precisely from these notes, one may discern in the thought of the later Merleau-Ponty a decisive reworking of the concepts of time, space, and movement, and thus a radically renewed approach to the notions of Nature, Human Being, and Being itself.

The key themes of this ontological itinerary emerge from the three intense confrontations the French philosopher maintains with the science, art, and psychology of his time. Indeed, it is precisely through these three apparent *détours* that Merleau-Ponty seeks to unsettle the formidable architecture of Western culture, an architecture grounded entirely in the principle of the *Kosmotheorós*. A crucial wager, to be sure – one that becomes, for the philosopher, the real "task of philosophy."[4]

> What exactly are we looking for when we try to abstract from nature the categories of substance, accident, potentiality, act, object, subject, in-itself, for-itself, which are traditionally involved in ontology? What can be the relation between the new ontology and classical metaphysics? Could it be the negation and the end of philosophy, or, on the contrary, is it perhaps the very same inquiry restored to its vital sources? With Hegel something cornes to an end. After Hegel, there is a philosophical void. This is not to say that there bas been a lack of thinkers or of geniuses, but that Marx, Kierkegaard, and Nietzsche start from a denial of philosophy. We might say that with the latter we enter an age of non-philosophy? But perhaps such a destruction of philosophy constitutes its very realization? Perhaps it preserves the essence of philosophy, and it may be, as Husserl wrote, that philosophy is reborn from its ashes?[5]

---

visible et l'invisible. Here the language, unlike that of the Working Notes appended to the posthumous manuscript by Lefort, is particularly concise, immediate, at times even more cryptic.

4 Merleau-Ponty, *In Praise of Philosophy and Other Essays*, quot., p.180.

5 Therein, pp-167-168.

It is to such questions that Merleau-Ponty intends to respond. A response capable of filling the post-Hegelian "philosophical void," a response that may go beyond and at the same time remain on this side of the "negation of metaphysics" carried out by Marx, Kierkegaard, and Nietzsche – a negation which "does not replace philosophy," and which, moreover, does not even rise to the level of its own crisis. Thus, the French philosopher continues, "we shall not find the answer to these questions by following the history of thought since Hegel"[6], it is, rather, a matter of bringing them back to their phenomenological origin; in other words, of freeing the notions of essence/existence, subject/object, possible/actual, spirit/matter from their subordination to the Absolute, and of considering them instead as "variants of the same world."[7]

From this point onward, an exacting work of dismantling the traditional philosophical categories begins – an undertaking that Merleau-Ponty, following in Husserl's path, intends to pursue. From *Phenomenology of Perception* onward, he attempts to retrace his steps toward the ground that founds every form of knowledge:

> It is a matter of describing, not of explaining or analysing. […] All my knowledge of the world, even my scientific knowledge, is gained from my own particular point of view, or from some experience of the world without which the symbols of science would be meaningless. The whole universe of science is built upon the world as directly experienced, and if we want to subject science itself to rigorous scrutiny and arrive at a precise assessment of its meaning and scope, we must begin by reawakening the basic experience of the world of which science is the second-order expression. Science has not and never will have, by its nature, the same significance *qua* form of being as the world which we perceive, for the simple reason that it is a rationale or explanation of that world.[8]

6 Ibid.
7 Cf. Merleau-Ponty, M., *The Visible and the Invisible*, quot., p. 41.
8 Id., *Phenomenology of Perception*, quot., p. IX.

This, in brief, is Merleau-Ponty's program: first, through a radical reconfiguration of the very notion of knowledge, to return to the lived experience of the world, to the ontological rehabilitation of the sensible; second, to inquire into the *how* by which the physiognomy of reality comes to be, starting from perception; and finally, to recognize the system of "anonymous functions" that configure the "style" of the world. The interrogation of Nature is thus the red thread that traverses Merleau-Ponty's entire reflection, the ultimate mark of how he unfolds the "task of philosophy," now rewritten in the terms of an inextricable implication of phenomenology, ontology, and *Gestaltpsychology*[9]. It is this implication whose fundamental outlines the French thinker traces above all in *The Visible and the Invisible* and in the set of working notes to which reference was made at the outset. Here it becomes the privileged point of orientation that allows the philosopher to pursue his search for the originary, for *Être brut* or *sauvage*; it constitutes the active background that opens access to "the vertical," that dimension through which it becomes possible to move beyond the impasse of deterministic and causalistic continuism. And it is here that the critique of Galilean–Newtonian science arises, in all its force and severity. The "passage to the vertical," the culminating outcome of Merleau-Ponty's later thought, is therefore to be understood as the moment in which classically conceived space and time implode. This means not only a de-substantialization of the "here" and the "now," but also a radical calling into question of our very existence. For what is at stake is not the defeat of Hegelian consciousness as it appears in "sense-certainty," but the collapse of the truth-regime of Western ontology – an ontology instituted by the two-thousand-year-old Platonic metaphysical tradition upon which the entire project of the

9 See at this point: Petronella Foultier A., *Incarnated meaning and the notion of Gestalt in Merleau-Ponty's phenomenology,* in "Chiasmi International", n. 17, 2015, pp. 53-75.

mathematization of the world has been erected. Merleau-Ponty – so the *Courses on Nature* delivered at the Collège de France from 1956 to 1960 attest – intends, in the wake of the scientific revolution of the twentieth century and its determination to strip both causalistic and finalistic conceptions of Nature of their meaning, to take leave of the so-called "natural laws" that govern the world according to the universal and necessary principle internal to them. These "universal natural laws," in fact, are anything but "laws of nature" in the sense of laws *belonging to* nature; rather, they are laws that scientific knowledge has inserted into nature from without – thereby concealing, in truth, the non-natural character of experimental procedures. Nature becomes the Object of analysis for the "Subject–Kosmotheorós", the absolute spectator. All this stands far indeed from Merleau-Ponty's own vision: "to think and to speak Nature without explaining it"[10]: This is the task – and the difficulty – of philosophy. A bold task, to be sure; yet one that is nonetheless urgent and of the first order, if only because the presumed objectivity of the natural sciences has shattered, has been put out of play by science itself and by its own new discoveries; "this core of universality around which history was to organize itself bas disintegrated.(…) even to those in it who post up complete certitudes (…) is not just that the human world is illegible, nature itself has become explosive."[11] Of this Nature, classical science cannot give an

10 Carbone, M. *Presentazione*, in Merleau-Ponty, M. *La Natura*, Bompiani, Milan 1996 (my translation).

11 Merleau-Ponty, *In Praise of Philosophy and Other Essays*, quot., pp.170-171. See, in particular: Vanzago L., *Nature, negativity, event. The question of the emergence of the centre in the ontology of the flesh,* in "Chiasmi International", n. 11, 2009, pp. 171-183; Vanzago L., *Process and events of nature. Merleau-Ponty and Deluze as readers of Whitehead,* in "Chiasmi International", n. 10, 2008, pp. 227-250; Id., *Raw being and the darkness of nature. Merleau-Ponty's appropriation of Schelling,* in "Chiasmi International", n. 16, 2014, pp. 239-251; Id., *The flesh between. Some remarks on André Green's remarks on Merleau-Ponty'*

account, obstinate "custodian of the idea of the in-itself and of the sovereign overview,"[12] because "even though its thesis shows that subject and object are inseparable, it assumes an absolute spectator"[13]. The objectification of Nature has been challenged by the findings of quantum mechanics: "What for centuries had, in the eyes of men, possessed the solidity of the earth now appears fragile; what was once our predestined horizon has now become a provisional perspective."[14] Nature, of which we are, withdraws from observation, eludes the visible. It bursts open before our very eyes; it "leaves a residue,"[15], unveils zones of shadow.

"Where am I?" and "What time is it?"[16]. Such questions have definitively lost their hold upon the world, their anchorage in the world. Zero-degrees or mere points of convergence, "lines of force" or "relational concordances", they are nonetheless "made of world." They constitute its only possible tangency. They reveal themselves – or better, they are visible – only as inflection points, to use a mathematical expression that, figuratively, suits well the chiasmatic movement: that is, as domains of individuation. But this means, upon closer view, describing them in their intimate contradiction: *"Where am I?"* and *"What time is it?"* are "inexhaustible" questions because "time and the place change continually"[17], because they refer to something else, to somewhere else that cannot itself be located in a defined "here" and "now." "The things – here, there, now, then – are no longer in themselves, in their

---

ontology and its relationship to psychoanalysis, in "Chiasmi International", n. 18, 2016, pp. 49-61.

12 Id, *Esquisse ontologique*, quot.

13 Cf. Ibid. (my translation).

14 Id., *In Praise of Philosophy and Other Essays*, quot., p.171.

15 Id., *Nature*, Northwestern University Press, Evanston, Ill. 2003, p. 83.

16 Id., *The Visible and the Invisible*, quot., p.105.

17 Therein, p.121.

own place, in their own time; they exist only at the end of those rays of spatiality and of temporality [...]. is no emplacement of space and time that would not be a variant of the others, as they are of it."[18] However, Merleau-Ponty explains that we are still at a superficial level of response in that "the question that arises here is not at bottom a question of knowing in what spot of a space taken as given, at what hour of a time taken as given, we are – but first what is this indestructible tie between us and hours and places, this perpetual taking of our bearings on the things, this continual installation among them, through which first it is necessary that I be at a time, at a place, whatever they be."[19] This is why, then, such questions are entirely "made of world"; why they attest to its only possible tangency; why, ultimately, they always express our implication in Being. "If we could scrutinize their ultimate motivation, we would find beneath the questions *where am* I? and *what time is it*? a secret knowledge of space and time as beings to be questioned, a secret knowledge of interrogation as the ultimate relation to Being and as an ontological organ."[20] Our implication in Being is therefore attested not by general laws of localization and temporalization, but by ephemeral, ungraspable "here" and "now." Ephemeral and ungraspable because, as already noted, they are always unmoored from themselves, always in a relation of divergence with respect to themselves. Everything said thus far finds further clarification precisely in *Esquisse ontologique*, where Merleau-Ponty specifies:

> Temporal distance = as a divergence in relation to a zero point of temporal distance. Time is not the sum of these divergences or modifications, but their tacit principle, that which makes them divergences *with respect to...* – that is, the opening of the "before us" and the "behind us," and of the circulation between the two. One should not even say *ecstasis*, which presupposes

18 Therein, p.114.
19 Therein, p.121.
20 Ibid.

> a point of departure in ipseity – the symbolic or animal matrix-time of "moments." Thus not serial time, but an open system as well: simultaneity is only a horizon, is only perceptual – time has frayed edges.[21]

The philosopher sketches an image of time as movement and possibility: the continuum is nothing more than what belongs to the order of the perceived; in truth, there are only "quanta" of time – of a time that can no longer be understood as the Memory of the world, its substrate or foundation. Rather, it is the simultaneous *Offenheit* of future and past and of the transgression of one into the other.

Transgression is, strictly speaking, an unforeseen passage, the possible movement that institutes the present, the "now," or better, the "nows." But if, as stated earlier, the here and the now are nothing other than our implication in Being, our tangency to the world, then it becomes entirely evident that such tangency is itself only a possibility. A possibility upon which the very articulative process of the archaeological and the teleological is literally suspended: origin and end are nothing but the figuration of the chiasmatic movement itself; they express its inner diplopia, its doubling and reciprocal inversion, like images in a mirror.

In fact, then, *arché* and *télos* "function" only within that unforeseen, possible passage that is transgression; they are – so to speak – the modes through which the world configures itself; they are visible/perceptible simply as local metamorphoses. There remains, therefore, nothing but to describe their movement – the *moved*, Merleau-Ponty would say – of implication, of refraction, of retraction. And here, just as in impressionist painting, movement is nothing other than allusive space-time, a lateral, indirect presentation of the world: the invisible that always transgresses, that overtakes and exceeds itself into/over the visible. Precisely because it is

21 Id., *Esquisse ontologique*, quot., (my translation)

never graspable in a here, in the staticity of Euclidean space, movement – inasmuch as it is the curvature of that space (and once again we are returned to the figure of the chiasm) – is the point of variance of the world, its possibility of *Offenheit* and, at the same time, its *il y a*, that is to say, its very excess.[22] In the Impressionist painting previously alluded to, Merleau-Ponty finds a precise point of reference for the sketching of his new ontology: lights, colors, reflections, large scale, in the movement and variations of the canvas, surround us. One cannot be a mere spectator. We are encompassed, encircled. If we fix our gaze on one part of a panel, the parts we do not see are acting within us as well. Monet's starting point is the little pond in his garden at Giverny. In painting, the small mirror of water becomes a cosmos; the visual absence of the sky transforms into an endless expansion of representation; the lights and colors of the water lilies and willows are timeless. "The "World" is this whole where each "part," when one takes it for itself, suddenly opens unlimited dimensions – becomes a *total part.*"[23] Thus, in the end, what there is is simply a "total part," an "îlot *d'esprit*", as Merleau-Ponty notes in the *Esquisse ontologique* – one of the possible openings of the world. And it is precisely from here that "that philosophy takes its start, namely, in a "there exists," in an "opening" toward "something," toward "that which is not nothing." It is this pre-objective Being, between the inert essence or *quidditas* and the individual localized at a point of space-time, that is the proper theme of philosophy. It can be said of this Being-the rose, said Angelus Silesius, is "without any why," "blooms because it blooms", the "rose-spectacle", the "rose-totality" – that it has no cause outside of itself, and moreover is not the cause of itself, that it is without ground, being the absence in

22 On the concept of "excess," see in particular Carbone, M., *Il sensibile e l'eccedente. Mondo estetico, arte, pensiero*, Guerini e Associati, Milan 1996.

23 Merleau-Ponty, M., *The Visible and the Invisible*, quot., p. 218.

principle of any ground."[24] Philosophy is reduced to silence if one seeks in it the direct expression of the "fundamental." This was precisely the error of Heidegger and Sartre: both had forbidden themselves all mirrors of Being, had annulled, by arresting the play of mirrors, the movement of Being itself. It is from this movement that the availability of the world occurs – in the sense that it is created – one might say: it is from here that its possibility, and therefore its very "there is," comes about. We are faced with a crucial juncture in Merleau-Ponty's thought. The task, the philosopher explains in his note dated September 23, 1958, is to succeed in thinking the *il y a*, the "there is," or rather excess, as possible – that is, as the very possibility of *Offenheit*. For what *there is*, the *il y a*, is simply "something," the *écart*, the "*modulation of the being in the world*"[25]. But what does all this mean? Is it not utterly contradictory to define the *there is*, the *déjà-là*, the act, the presence, as possible? Here the French philosopher continues his work of radically overturning the history of Western thought, founded upon the process of "rectification"[26] of reality instituted by Euclidean logic – and, once again, it is the very concept of three-dimensional space that is called into question:

> The Euclidean space is the model for perspectival being, it is a space without transcendence, positive, a network of straight lines, parallel among themselves or perpendicular according to the three dimensions, which sustains all the possible situations – Underlying appropriateness of this idea of space (and of velocity, movement, time) with the classical ontology of the *Ens realissimum,* of the infinite entity. The topological space, on the contrary, a milieu in which are circumscribed relations of proximity, of envelopment, etc. is the image of a being that, like Klee's touches of color, is at the same time older than everything and

24 Id., *In Praise of Philosophy and Other Essays*, quot., p. 178.
25 See Id., *The Visible and the Invisible,* in particular pp. 201-211
26 It is, in other words, the "straight-line" vision upon which Euclid based his "geometrization" of the world.

> 'of the first day' [...], that the regressive thought runs up against without being able to deduce it directly or indirectly [...] from Being by itself, that is a perpetual *residue*. It is encountered not only at the level of the physical world, but again it is constitutive of life, and finally it founds the *wild* principle of Logos. It is this wild or brute being that intervenes at all levels to overcome the problems of the classical ontology (mechanism, finalism, in every case: artificialism).[27]

Ultimately, one must "take topological space as the model of Being[28] – that is, the space that makes possible the unmotivated arising of brute Being"[29]. Here enters the shift from *why* – the expression of the search for a causal and necessary principle – to *how*, in the sense that this is not about asserting the priority of potentiality over actuality – such a move would not free Merleau-Ponty from the dichotomous impasse – but rather about conceiving the very movement of one upon the other, their reciprocal exchange, their essential bifurcation, their simultaneous enveloping. This is what constitutes the "modulation of the being in the world" previously referred to, which transforms space, the *there is*, into movement – that is, into passage, into *être de champ*. "What would space be – the philosopher asks – without movement? Not merely undefined, but null, for it is movement that constructs the very framework of space"[30]. We are then speaking of movement as that "something" which sustains – or rather, which allows – the emergence of space. And here the reference to *The Perception of Casuality* by Albert Michotte[31] is quite clear, to which Merleau-Ponty constantly refers in these notes. Not only that. The philosopher's reference to the pictorial expression

27 Merleau-Ponty, M., *The Visible and the Invisible,* quot., pp.210-211.
28 Ibid.
29 Ibid.
30 Id., *Working Note of September 23, 1958*, quot., (my translation)
31 Michotte, A., *The Perception of Causality*. Routledge, London 2017.

of Monet, Cézanne, and Klee becomes even clearer, where movement qualifies the space of the canvas, its *there is*, as its "possibility of.".

If we delve deeper into the matter and link movement thus understood to the figure of the chiasm, we immediately realize that it is nothing other than the possible act, the locus of curvature that opens access to Being. We can therefore speak of time and space only starting from that articulatory process – movement itself – which always fails to individuate them or to "point them out," for such indication is constantly anticipated or surpassed, always in excess of itself. Thus, the "heres" and the "nows" are from the outset irretrievably lost, and yet they fill our places of existence. Certainly, these are mortal places, never secured, never anchored to the certainty of necessity. From this arises a conception of Being, and thus of Nature, that is not substantialist, but always suspended upon the obscure principle of becoming, of its possibility.

> Nature – explains Merleau-Ponty – possesses a temporal character; we only ever grasp it in its manifestations [...] just as the wave is only an encroachment, so too is Nature an encroachment of serial time and space. Its sound is likewise a nominiform being, nott-localizable in a series of instantaneous beings but only in the process of each of them.[32]

Similarly, the philosopher continues,

> the electron is not found where its charge is. The electron is a certain property that plays a focal role, and to which we relate the observed physical events in a "situation current." [...] the electron does not exist in the sense of absolute Being, which is all or nothing: the electron does not reside in a punctual and objective spatiotemporality; it is an "ingredient" (this word also has the sense of making an ingression) in its whole vicinity, it is the hallway of certain "traces," of certain "roles" observed by the observer. (...) Bui (...) Whitehead pushes back the notions of

32 Merleau-Ponty, M., *Nature*, quot., p. 120.

> action-at-a-distance and action-by-transinission, (…) He speaks of "overlapping relations," of "relations of extension," to be understood before all spatiotemporal specification as the foundation of time and space as well as their union. The spatiotemporal unities overlap. The task imposed on the philosophy of Nature would be lo deepen the relation that exists between these unities. (…) The object is the abridged way of marking that there was an ensemble of relations.[33]

The preference for wave mechanics[34], a system in which time, velocity, and space are mutually dependent and directly influenced by the external observer – finds once again its strongest support in a passage from *The Perception of Casuality* by Albert Michotte, to which Merleau-Ponty refers in the *Esquisse ontologique*: "the physical 'world' in question here is no longer the world of physical science, as revealed by measuring instruments; it is the world of things, as it appears to the subject on simple inspection, his 'phenomenal world', disclosed in this case by the indications which he gives as a human 'recording instrument'"[35]; causality, then, is nothing more than a particular experience of the relationship between two objects or events. At this level, Merleau-Ponty explains, "The critique of unique emplacement makes us understand the ontological value of perception. What I perceive is both for me and in the things. Perception is made starting from the interior of Nature: I become aware of the lamp which is held in front of me in a temporal sense."[36]Michotte's phenomenological analysis demonstrates incontrovertibly that the impression of causation[37] is an immediate perceptual datum; it is the expressive quality of movements. Therefore, rather than

33 Cf., therein, pp. 115-116.

34 Cf., Id., *Esquisse ontologique*, quot.

35 Michotte, A., *The Perception of Causality*, quot., p. 306 and Merleau-Ponty, M., *Esquisse ontologique*, quot.

36 Merleau-Ponty, M., *Nature*, quot., p. 117.

37 Embree L., *The impression of causality: Merleau-Ponty on Michotte,* in "Chiasmi International", n. 11, 2009, pp. 311-320.

speaking of causal links, it is more appropriate to understand it as the "segregative" influence of objects upon movement. Here, perception and relation say the same thing, expressing their mutual "ingredient," their irreducible articulation. In this same vein, moreover, Merleau-Ponty's critique of reflective philosophy should be read, for by substituting the possible with actualism, it fails to account for the unreflected, the "edges of nature (that) are always ragged"[38] but yet it presumptively situates itself within the One.[39] From this arises the danger highlighted by the French thinker: a speculative conception that a priori grounds access to Being, encloses the sign within meaning, Sainte-Victoire within its canvas, the movement of color within the fixity of white space. It is therefore necessary – insists Merleau-Ponty – to

> replace the notions of concept, idea, mind, representation with the notions of *dimensions,* articulation, level, hinges, pivots, configuration. The point of departure = the critique of the usual conception of the *thing* and its *properties* → critique of the logical notion of the subject, and of logical inherence → critique of the *positive* signification (differences between significations), signification as a separation *(ecart),* theory of predication – founded on this diacritical conception[40].

Hence the problem that Merleau-Ponty, in the *Esquisse ontologique*, places at the center: we believe in a κόινος κόσμος and we only ever have a ἴδιος κόσμος"[41]. And it is precisely the latter that we must deal with. The "conviction [...] of an κόινος κόσμος that integrates everything" in fact, fails to explain the sense of 'explosive nature'; the necessary and universal 'generality' within which the order of the natural world had been mapped does not reach 'our private spectacle.'[42]. What

38 Merleau-Ponty, M., *Nature*, quot., p. 114.
39 Cf. Id., *Esquisse ontologique*, quot.
40 Id., The Visible and the Invisible, quot., p.224.
41 Id., *Esquisse ontologique*, quot., (my translation)
42 Ibid. (my translation).

"there is" is simply "an unforeseen communication from one *îlot d'esprit* to another"[43], as if to say that every ἴδιος κόσμος stands as a possible, eventual, "virtual" opening of the world. Our implication in Being, the philosopher emphasizes here, is far from being founded on the guarantee of a universal law – precisely that reassuring and ideal law of a "κόινος κόσμος that integrates everything[44]. It is instead constantly determined by parts, singularities, rays of the world. The result is a vision of the world that is no longer singular, but rather perspectival, lateral, partial: an individual cosmos, an ἴδιος κόσμος indeed, which cancels the truth of the universe and ventures into the multiverse – or rather, the *pluriverse* of truths. What reflexive philosophy lacks = that which lies between the views. *Welt*, *Mitwelt*, or truth in difference. A world truly with multiple entrances."[45] But, to succeed in conceiving the full sense of the 'ἴδιος κόσμος, of the world *à plusieurs entrées*, a new *Befragung* is necessary, Merleau-Ponty continues – one founded on a *Besinnung* no longer understood as "reflection on the world or analysis of the world," but as a descent toward meaning (the world in flesh and bone), the unveiling of perceptual meaning, a meaning that is at once latent."[46] "The world – the philosopher emphasizes again – is individual, of an absolute individuality, prior to the singular and the plural."[47] The full meaning of this expression takes us back to the final indications of the note *Pour l'ontologie*, where – once again, as if to underscore their intimate cohesion – the philosophical and artistic vocabularies appear fused: "*The something is a style.*" Everything else can be connected to it, and it is there simultaneously as a specific element and as a measure. In this sense, it is a world, that is, a total part. Reciprocally, the world is the *something of something.*

43 Ibid. (my translation).
44 Ibid. (my translation).
45 Ibid. (my translation).
46 Ibid. (my translation).
47 Ibid. (my translation).

The world as the sum of specific elements and the world as style are inseparably linked: the world is the common style of the specific elements, their logic, and their connection."[48] One must speak, as is evident from the brief passages of the *Esquisse ontologique* concerning the inseparable correlation between the ἴδιος κόσμος and the κόινος κόσμος of a doubling of one upon the other: this will be the task of every philosophy. Indeed, "What is always lacking is a world with multiple entrances: either there is one, or there is an infinity. There are not several (which means a greater… *Offenheit*)."[49] To speak of a world *à plusieurs entrées* is therefore to understand in what sense a *Befragung* takes place upon the world, in what sense a *Besinnung* of the world is attained. There is no single vision of the world, nor are there infinite visions of the world: the "modalities of total existence," mentioned at the outset, are multiple and, at the same time, absolutely possible. Just as we cannot apprehend all the facets of the object before us – even if it seems to present itself to us in its entirety, even if we depict it thanks to the seams formed around its multiple profiles – so too we cannot grasp the totality of the world, since it presents itself not to a thesis, to a *Vorstellen*, or to representational knowledge, but to an opening, to a perception–non-perception, to a multiplicity which, like the visible, always promises more to be seen. The "*vertu du proche*" and the "*totalité de proximité*" that Merleau-Ponty discusses in his *Notes de lecture*[50] on Rudolf Arnheim's *Art and visual perception*[51] thus become the privileged expressions of our contact with the world; a contact which, precisely because it is near, attainable, and possible, is

48 Id., *Pour l'ontologie*, unpublished note dated September 23, 1958, quot.,

49 Cf. Id., *Esquisse ontologique*, quot., (my translation).

50 Cfr. Id., *Notes de lecture: R. Arnheim,* Art and visual perception*, a psychology of the creative eye* (Boîte V, D.M.O., B.N.F.).)

51 Arnheim, R., *Art and visual perception, a psychology of the creative eye*, University of California Press, Berkeley and Los Angeles 1954.

always configured through its very withdrawal, through an absence that allows its silent presence to appear. Here a reversal takes place: it is absence, it is the invisible that speaks, that opens toward presence, toward the visible, the space of the λέκτον, of the sayable; of a sayable which, precisely because it arises as absence, takes shape as silence *secundo*, insofar as it has already been frequented. In the same way, one must conceive of the *Befragen*: a simple opening to the universe of language, to that familiarity with silence which returns as word. And this is nothing other than the sole mode of attending to Being, of grasping its proximal distance, the lateral relations of *Ineinander*. In this sense, *Besinnung* becomes, at this level, a *Rückfrage*, an originary questioning: "One abandons," we read in the *Esquisse ontologique*, "the said (*das Fertige*) in order to seek the sayable; this does not mean coinciding with it, but bringing it to expression, that is, designating it univocally through available significations. Thus the diplopia of *Besinnung*; it returns backward, but also creates, moving toward the anterior pole of language."[52] All of this is further clarified in a passage from *Notes pour choses à faire*, where Merleau-Ponty emphasizes that philosophy – understood as interrogation – remains a philosophy with a single entrance, originating from the knowing subject. Yet, as soon as it opens onto existence, onto the lived experience, it doubles or truly multiplies. Philosophy is precisely the "consciousness" of this "explosion"; or rather, the consciousness of a "membrane" that makes it possible. Thus, what there is, the *il y a*, is simply the presence of these "converging gleams."[53] Thus, it is only by replacing reflective "consciousness" with *Befragung–Besinnung* that it becomes possible to speak of interrogative ontology in the sense of the "vision of the nature-human-

52 Merleau-Ponty, M., *Esquisse ontologique*, quot., (my translation).
53 Id., *Notes pour choses à faire*, (Boîte III, D.M.O., B.N.F.) (my translation).

being"[54]. In his note dated September 23, 1958, Merleau-Ponty makes explicit the character of interrogative ontology: it consists in considering the value of notions such as essence, existence, subject, object, matter, possible, and actual as so many modes of being, rather than thinking from the standpoint of any one of them. This is because all these modes of being share their belonging to a single world. It is therefore a matter of understanding how these modes of being articulate themselves within Being; which means, once again, taking leave of the logical-rational standpoint that always explains the particular through the universal and fails to account for the crisis of all our "categories." The notions of Nature[55], Man, and God indeed reveal a passage, a transgression of one into the others, a confusion that attests to their connection and reciprocal implication, that highlights their solidarity within Being, and that finally makes visible the necessity of a radical investigation of the "framing of Being."[56] This investigation brings us back to the *Notes de lecture sur Arnheim*, precisely where Merleau-Ponty, returning to the theme of perception – a theme he never abandoned – emphasizes that it is configured as an "anonymous function" (perception is always in the world of the *one*—one says, one thinks, etc.) through which it becomes possible for us to grasp the "framework" of Being. But this possibility is constantly deferred, since the "perceptive *how*" is always in excess of the possible, and Being, its framework, is given to us only from a perceptual "angle." An angle of the world, then,

---

54 Id., *Working note dated October 7, 1958,* quot., (my translation).

55 On the concept of Nature in Merleau-Ponty see, in particular: Foti V.M., *Espression, alterity, and the phylosophy of nature in Merleau-Ponty's dialogue with the rationalists,* in "Chiasmi International", n. 11, 2009, pp. 279-291; Larison, M., *Autor du concept de nature dans le dernier Merleau-Ponty,* in "Chiasmi International", n. 7, 2005, pp. 393-405; Umbelino, L.A., *La nature indirecte et les couleurs de l'invisible. Notes à propos du projet ontologique de Merleau-Ponty,* in "Chiasmi International", n. 12, 2010, pp. 273-285.

56 Merleau-Ponty, M., *Signs*, quot., p. 181.

which allows the philosopher to affirm that "the Offenheit of the Umwelt and the openness of the perceiver are the same thing"[57]. To the question "Where is the thing? Where is Mount Sainte Victoire?", Merleau-Ponty adds in a brief marginal note to *L'interrogation de l'Être brut*: "There is no defined *where*. The *where* is always anticipated or surpassed; it can never be grasped at a single point."[58] And again, in the *Notes pour la suite*, the philosopher emphasizes the necessity that there always be a "diaphragm." This is certainly not a "fixed diaphragm," but rather a "transcendent diaphragm," which, once a certain point is surpassed, carries within itself its own "disintegration," its own "dedifferentiation."[59] How to say that the inflectional curvature discussed at the outset is, from the very beginning, a possible opening of the world. Yet it is precisely the sense of this possibility that must be interrogated, if only because we ourselves are part of it – also, therefore, as the *Offenheit* of which we are, constitutively, "torn to shreds" (in the words of Merleau-Ponty himself).[60]. This is why shadow, relief, edge, and distance cannot be considered mere metaphors, but as *monde en plein*, or rather, "total parts of the world," "indirect significations," dialectical poles of the same linguistic–sign system, "masks of the Same."

The fragmentary outcome of Merleau-Ponty's trajectory, recognizable in his often allusive, metaphorical, almost literary language, eludes any claim to conclusion and instead relies upon a word that withdraws, silent, into what the philosopher himself called "the labyrinths of ontology." Yet it is precisely in the withdrawing of the word, in the contraction of its space, that its possible opening is given – through the surpassing of

57 Id., *Esquisse ontologique*, quot., (my translation).
58 Cf., Ibid. (my translation).
59 Cf. Id., *Notes pour la suite,* in *Notes pour choses à faire,* quot., (my translation).
60 Cf. Id., *Esquisse ontologique*, quot., and *Working Note of September 23, 1958*, quot., (my translation).

philosophical interrogation, which therefore begins anew.[61] If this is indeed the sense of our being-in-the-world, here perhaps lies the meaning of our very living, insofar as it can be understood only as movement, the possible "there is" of/between birth and death.

61 Cf. in this regard Barbaras, R., *Le dédoublement de l'originaire*, in M. Merleau-Ponty, *Notes de cours sur L'origine de la géométrie de Husserl*, followed by *Recherches sur la phénoménologie de M. Merleau-Ponty*, under the direction of R. Barbaras, P.U.F., Paris 1998; pp. 289-303.

BIBLIOGRAPHY

Alcofarado D., *M. Merleau-Ponty. A pintura como interrogaçao radical do mundo*, "Revista Fac. Letras y Filosofia", n. 7, 1990, pp. 215-258.

Ales Bello A., *"Essere grezzo" e hyletica fenomenologica. L'eredità filosofica del visibile e l'invisibile,* in "Chiasmi International", n. 10, 2008, pp. 139-161.

Alloa E., *La chair comme diacritique incarné,* in "Chiasmi International", n. 11, 2009, pp. 249-262.

Alloa E., *Merleau-Ponty à Madagascar. L'épreuve de l'étranger et la décolonisation de la pensée*, in "Chiasmi International", n. 19, 2018, pp. 115–127.

Alloa E., Revel, J., *Merleau-Ponty politico* in "Chiasmi International", n. 19, 2018.

Almansi A, *Di alcuni temi della critica merleau-pontyana*, "Segni e comprensione" I, 1, 1987, pp. 55-68.

Angelet B., *L'œil et l'esprit: "avoir toute la vie devant soi"*, "Krtiek", 7, 1984 pp. 47-86.

Arnheim R., *Art and visual perception, a psychology of the creative eye*, University of California Press, Berkeley and Los Angeles 1954.

Aronson R., *Vicissitudes of the Dialectic from Merleau-Ponty's* Les Aventures de la dialectique *to Sartre's Second Critique*, "Philosophical Forum", XVIII, 4, 1987 pp. 358-391.

Bannan J.F., *Merleau-Ponty and Metaphor,* "The Review of Metaphysics", 47, 2, 1993 p. 370.

Barbaras R., *Phénoménalité et signification dans* Le visible et l'invisible, "Les Cahiers de Philosophie", 7, 1989, pp. 25-53.

Barbaras R., *De l'être du phénomène. Sur l'ontologie de Merleau-Ponty*, Millon, Grenoble. 1991.

Barbaras R., *Le sens de l'auto-affection chez Micbel Henry et Merleau-Ponty*, "Epokhé", 2, 1991, pp. 91-111.

Barbaras R., *La perception. Essai sur le sensible*, Hatier, Paris 1994.

Barbaras R., *Le tournant de l'expérience: Merleau-Ponty et Bergson*, "Philosophie", 54, 1997, pp. 33-59.

Barbaras R., *Le tournant de l'expérience. Recherches sur la philosophie de Merleau-Ponty*, Vrin, Paris 1998.

Barbaras, R., *Le dédoublement de l'originaire*, in Merleau-Ponty, M., *Notes de cours sur L'origine de la géométrie de Husserl*, followed by *Recherches sur la phénoménologie de M. Merleau-Ponty*, under the direction of R. Barbaras, P.U.F., Paris 1998; pp. 289-303.

Barbaras, R., *Le tournant de l'expérience. Recherches sur la philosophie de Merleau-Ponty*, Vrin, Paris 1998.

Barbaras R., *Le désir et la distance. Introduction à une phénoménologie de la perception*, Vrin, Paris 1999.

Barbaras R., *Merleau-Ponty aux limites de la phénoménologie,* in "Chiasmi International", n. 1, 1999, pp. 199-211.

Barbaras R., *Les trois sens de la chair. Sur une impasse de l'ontologie de Merleau-Ponty,* in "Chiasmi International", n. 10, 2008, pp. 19-33.

Barbaras R., *Phenomenology and the poetic,* in "Chiasmi International", n. 17, 2015, pp. 189-199.

Barral M.R, *La metafisica nel pensiero di Merleau-Ponty,* in B, D'amore, A. Ales Bello (a cura di), *Metafisica e scienze dell'uomo*, Borla, Roma 1982.

Barrett C., *Merleau-Ponty and the Phenomenology of Perception,* "Philosophy", 1987, pp. 123-139.

Barry J., *The Physics of Modern Perception. Beyond Body and World,* [Heidegger-Merleau-Ponty], "The Journal of Speculative Philosophy", 4, 4, 1990, pp. 287-297.

Barry J., *The Technical Body: Incorporating Technology and Flesh,* Philosophy Today, 35, 4, 1991, p. 390.

Bayer, R., *Merleau-Ponty et l'existentialisme,* "Revue Philosophique de la France et de L'Etranger", n. 87, 1962.

Bazzanella E., *Orizzonte, passività e soggetto in Husserl e Merleau-Ponty*, Guerini Scientifica, Milano 1995.

Bech, J. M., *'Institution' at the Crossroad: Ontological Extension or Corporeal Restatement,* in "Chiasmi International", 11, 2009, pp. 201–247.

Bello A., *De Sartre à Merleau-Ponty, Dialectica de la libertad y el sentido*, Publicaciones Universidad de Murcia, Murcia 1979.

Bello A., *Self and other. Communication and Love*, "Review of Existential Psychology & Psychiatry", 18, 1982-83, pp. 155-180.

Bello A., *The Body in Interpersonal Relations: Merleau-Ponty,*

University press of America, Lanham; London 1984.
Bello A., *Texts and Dialogues*, “Choice”, n. 10, 1992, p. 1560.
Bello A., *Merleau-Ponty vivant*, “Choice”, n. 29, 7, 1992, p. 1092.
Belot D., *“Cette idéè d’une histoire comme champ…” des aventures de la dialectique à la préface de signes,* in “Chiasmi International”, n. 19, 2018, pp. 57-68.
Bergson, H., *Matter and Memory*, transl. by N. M. Paul and W. Scott Palmer, Allen Unwin, London 1962.
Bernet R., *The Phenomenon of the Gaze in Merleau-Ponty and Lacan,* in “Chiasmi International”, n. 1, 1999, pp. 105-119.
Bernet R., *Philosophy and literature – literature and philosophy,* in “Chiasmi International”, n. 19, 2018, pp. 255-271.
Bertolini P., *Fenomenologia e pedagogia*, Malipiero, Bologna, 1958.
Bertolini P., *L’esistere pedagogico., Ragioni e limiti di una pedagogia come scienza fenomenologicamente fondata*, La Nuova Italia, Milano-Firenze 1988.
Bertolini P., *Pedagogia fenomenologica. Genesi, sviluppo, orizzonti*, La Nuova Italia, Firenze 2001.
Boisseau D., *Merleau-Ponty et l’ambiguité*, “Etudes Philosophiques”, n. 2, 1992, pp. 229-243.
Bonomi A., *Esistenza e struttura. Saggio su Merleau-Ponty*, II Saggiatore, Milano, 1967.
Borg J.L., *Le marxisme dans la philosophie socio-politique de Merleau-Ponty*, “Revue Philosophique de Louvain”, 73, 1975, pp. 481-509.
Bourgeois P.L., *The Existential and Temporal Self Merleau-Ponty*, “Phenomenological Inquiry”, 13, 1989, pp. 64-81.
Bourgeois P.L., *The Integration of Merleau-Ponty’s Philosophy*, “Southwest Philosophy Review”, n. 5, 1989, pp. 37-50.
Bourgeois P.L., *Semeiology and the Phenomenology of Language. Merleau-Ponty and Ricoeur*, “Philosophy Today”, 15, 1991, pp. 162-171.
Bourgeois P.L., *Merleau-Ponty and Heidegger. The Intentionality of Transcendence, the Being of Intentionality*, “The Journal of the British Society for Phenomenology”, 25, 1, 1994, pp. 27-33.
Bourgeois P.L., *From Common Roots to a Broader Vision: A Pragmatic Thrust of European Phenomenology*, “American Catholic Philosophical Quarterly”, 70, 3, 1996, p. 381.
Brand G. 1955, *Mondo, Io e Tempo nei manoscritti inediti di Husserl*, Bompiani, Milano 1960.
Brena G.L., *La struttura della percezione. Studio su Merleau-Ponty*,

Vita e Pensiero, Milano 1969.
Brena, G. L., *Alla ricerca del marxismo: Maurice Merleau-Ponty*, Dedalo, Bari 1977.
Brena G.L., *Ateo per malinteso. La religione secondo M. Merleau-Ponty*, "Fenomenologia e Società", XVII, 3, 1994, pp. 60-83.
Bucur D., *Le rapport à autrui comme structure (Gestalt),* in "Chiasmi International", n. 17, 2015, pp. 77-93.
Burke P., *Il potere morale del volto del bambino*, "Chiasmi International", n. 1, 1999, pp. 137-152.
Busch T.W., *Ethics and Ontology. Levinas and Merleau-Ponty*, "Man World", n. 25, 2, 1992, pp. 195-202.
Cabral I., *Maurice Merleau-Ponty's Logic of Ambiguity*, "Analecta Husserliana", n. 29, 1990, pp. 323-337.
Cacciari M., *Geo-filosofia dell'Europa*, Adelphi, Milano 1994.
Caeymaex, F., Cormann, G., *Sartre/Merleau-Ponty, andate e ritorni*, in "aut aut", 381, 2019, pp. 120–148.
Calabrò D., *Maurice Merleau-Ponty: il tracciato storico di un'ontologia del linguaggio,* "Criterio", X, 1992, pp. 248-253.
Calabrò D., *Sartre e Merleau-Ponty: analisi di una polemica,* "Atti dell'Accademia Peloritana dei Pericolanti", LXX, 1994, pp. 163-195.
Calabrò D., *Merleau-Ponty interprete di Hegel*, "Atti dell'Accademia Peloritana dei Pericolanti", LXXI, 1995, pp. 329-362.
Calabrò D., *Tempo vissuto e temporalità originaria in Maurice Merleau-Ponty*, in *Il concetto di tempo*, a cura di G. Casertano, Atti del XXXII Convegno nazionale della SFI, Loffredo, Napoli, 1997, pp. 263-271.
Calabrò D., *Maurice Merleau-Ponty e il "labirinto dell'ontologia"*, in "Chiasmi International", n. 1, 1999, pp. 153-167.
Calabrò D., *L'infanzia della filosofia. Saggio sulla filosofia dell'educazione di Maurice Merleau-Ponty*, UTET, Torino 2002.
Calabrò, D., *The Correspondance. Sartre and Merleau-Ponty*, Mimesis International, London–Milan 2018.
Calvino I., *Palomar*, Einaudi, Torino 1983.
Carbone M., *L'occhio e il corpo. Merleau-Ponty tra universo pittorico e universo televisivo*, in VV.AA., *Problemi del senso*, Quaderni di Studi Mantovani, Mantova, 1982, pp. 103-116.
Carbone M., *La forma precaria del linguaggio in Merleau-Ponty. Una lettura di "La prose du monde"*, "Fenomenologia e scienze dell'uomo", n. 2, 1982, pp.137-145.
Carbone M., *Tempo e parola. Merleau-Ponty lettore di Proust*, "Segni e comprensione", I, 1, 1987, pp. 7-23.

Carbone M., *Il sensibile e il desiderio. Merleau-Ponty, Lyotard e la pittura, aut-aut*, 232-233, 1989, pp. 63-86.
Carbone M., *Ai confini dell'esprimibile. Merleau-Ponty a partire da Cézanne e da Proust*, Guerini e Associati, Milano 1990.
Carbone, M. *Alla ricerca dell'*a-filosofia. Merleau-Ponty e la *Einleitung*, alla *Phänomenologie des Geistes*, in VV.AA., *Negli specchi dell'Essere. Saggi sulla filosofia di Merleau-Ponty*, Hestia, Como 1993.
Carbone M., *"Ce qui se voit en nous". La peinture et le narcissisme de l'Etre dans "L'Oeil et l'esprit"de Maurice Merleau-Ponty*, in *La voix des phénomènes. Contributions à une phénoménologie en hommage a Ghislaine Florival*, a cura di R. Braisart, R. Célis Publications des Facultés Universitaires Saint-Louis, Bruxelles 1995, pp. 213-227.
Carbone, M., *Il sensibile e l'eccedente. Mondo estetico, arte, pensiero*, Guerini e Associati, Milan 1996.
Carbone M., *Presentazione*, in M. Merleau-Ponty, *La Natura, Cortina*. Milano, 1996, pp. XI–XXII.
Carbone, M., *Mondo estetico, arte, pensiero*, Guerini e Associati, Milano 1996.
Carbone M., *Tracce di non. Variazioni sul tema dello stile*, "Chiasmi", 1, 1998, pp.97-102.
Carbone M., *Il tempo mitico delle idee*, "Chiasmi international", 1, 1999, pp. 213-231.
Carbone, M., *The Flesh of Images: Merleau-Ponty between Painting and Cinema*, SUNY Press, New York 2016.
Carbone M., Levin D., *La carne e la voce. In dialogo tra estetica ed etica*, Mimesis, Milano 2003.
Carcano Filiasi P., *Il bambino e gli altri*, Armando ed., Roma, 1968.
Cardim L.N., *Science et philosophie chez Merleau-Ponty,* in "Chiasmi International", n. 8, 2006, pp. 17-42.
Carron G., *De l'expérience à l' "événement": les enjeux de la pensée d'un "symbolisme originaire",* in "Chiasmi International", n. 13, 2011, pp. 481-497.
Carron G., *La virtù sans aucune résignation,* in "Chiasmi International", n. 16, 2014, pp. 41-54.
Casey E.S., *Visibilizing the invisible in painting,* in "Chiasmi International", n. 19, 2018, pp. 239-253.
Cassinari F., *Tra Heidegger e Sartre. L'ontologia del tempo-mondo in Maurice Merleau-Ponty*, "Fenomenologia e società", XVII, 3, 1994, pp. 16-31.
Cataldi S.L., *The Conception of Line in Heidegger and Merleau-*

Ponty, "Philosophy Today", n. 32, 1988, pp. 327-337.

Cataldi S.L., *Emotion, Depth and Flesh. A Study of Sensitive Space. Reflexion on Merleau-Ponty's Pbilosophy of Embodiment*, State University of New York Press, Albany 1993.

Cavalcanti M., *A intencionalidade segundo Maurice Merleau-Ponty,* PUC, Rio de Janeiro, 1978.

Cavaleri B., *Sguardi e parole: lo stupore, Il paesaggio della "nuova ontologia"*, "aut-aut", 232-233, 1989, pp. 43-62.

Cavarero A., *Corpo in figure, Filosofa e politica della corporeità*, Feltrinelli, Milano 1955.

Cefai D., *L'experience et l'expression. Essai sur la pensée de Maurice Merleau-Ponty*, "Canadian Journal of Political Science", n. 22, 2, 1989, pp. 445-446.

Certeau M. DE, *La folie de la vision*, "Esprit", n. 6, 1982, pp. 89-99.

Charcosset J.P., *Merleau-Ponty. Approches Phenoménologiques*, Hachette, Paris 1981.

Charcosset J.P., *La tentation du silence*, "Esprit", 6, 1982 pp. 53-63.

Chiesa P., *La sublimazione come cifra della reversibilità,* in "Chiasmi International", n. 11, 2009, pp. 157-170.

Chouraqui F., *Temporal thickness in Merleau-Ponty' notes of May 1959,* in "Chiasmi International", n. 13, 2011, pp. 407-427.

Chouraqui, F., *The Paradox of Power*, in "Chiasmi International", 19, 2018, pp. 69–85.

Clair A., *Merleau-Ponty lecteur et critique de Bergson: le statut bergsonien de I'intuition*, in "Archives de Philosophie", n. 59, 2, 1996, pp. 203-218.

Cocchi P., *Filosofa e impegno politico in Merleau-Ponty,* "Critica marxista", XXVI, 1, 1988, pp. 141-160.

Cohen R.A., *Merleau-Ponty, the Flesh and Foucaualt*, "Philosophy Today", n. 28, 1984, pp. 329-338.

Collins C., *Body-Intentionality*, "Inquiry", n. 31, 1988, pp. 495-518.

Compton J., *Sartre, Merleau-Ponty and the Human Freedom*, "The Journal of Philosophy", n. 79, 10, 1982, pp. 577-588.

Compton J., *Review of Maurice Merleau-Ponty. Between Phenomenology and Structuralism,* "History and Theory", 26, 3, 1987, pp. 365-373.

Compton J., *Merleau-Ponty's Metaphorical Philosophy*, "Research in Phenomenology", 23, 1994, pp. 221-26.

Cook D., *Writing Philosophy and Literature. Apology for Narcissism in Merleau-Ponty,* "Eidos", 4, 1985, pp. 1-9.

Coole D., *The Aesthetics Realm and tbe Lifeworld: Kant and Merleau-Ponty,* "History of Political Thought", n. 5, 1984, pp.

503-526

Costantino S., *Merleau-Ponty interprete di Hegel e di Marx*, "Segni e comprensione", I, 1, 1987, pp. 24-46.

Costantino, S. *La struttura del linguaggio in Merleau-Ponty,* Corso Editore, Ferrara 1976.

Cotroneo G., *Il dileguare della prima certezza*, in Aa. Vv., Il mondo incerto, Laterza, Roma-Bari, 1994, pp. 5-25.

D'ippolito B.M. *Ontologia e storia in E. Husserl*, Rumma, Salerno, 1968.

Dalla Vigna P., *Dal figurale al sensibile. Note sul rapporto tra Merleau-Ponty e Lyotard, a partire da un testo di Mauro Carbone*, "Chiasmi", 1, 1996, pp. 27-37.

Dalla Vigna P., *L'esistenzialismo francese e la sua autocritica. Merleau-Ponty e Sartre*, "Fenomenologia e Società", XVII, 1, 3, 1994, pp. 32-47.

Damish H., *Le versant de la parole*, "Bulletin de Psychologie", 236, XVIII, 1964, p. 105.

Dastur F., *Consciousness and Body in the Phenomenology of Merleau-Ponty. Some Remarks Concerning Flesh, Vision and World in the Late Philosophy of Maurice Merleau-Ponty*, "Analecta Husserliana", n. 17, 1984, pp. 117-125.

Dastur F., *La temporalité chez Merleau-Ponty (Merleau-Ponty entre Husserl et Heidegger)*, in G. Florival (a cura di), *Etudes d'anthropologie philosophique*, vol. *Dimension de l'exister*, Institut Superieur de Philosophie, Peeters, Louvain-la Neuve/ Paris 1994, pp. 19-32.

Dastur F., *Merleau-Ponty et Hegel. Ontologie et dialectique,* in "Chiasmi International", n. 11, 2009, pp. 33-47.

De Saint Ubert E., *Conscience et expression chez Merleau-Ponty: l'apport du cours inédit sur le monde sensible et le monde de l'expression,* in "Chiasmi International", n. 10, 2008, pp. 85-107.

De Saint Aubert E., *Les "différenciations d'une seule et massive adhésion a l'être qui est la chair". Étude de la notion de polymorphisme chez Merleau-Ponty,* in "Chiasmi International", n. 11, 2009, pp. 185-199.

De Wahelens, A., *Una filosofia dell'ambiguità*, Preface to *La struttura del comportamento*, Bompiani, Milano 1963.

Delaunay, R., *Du cubisme* à l'art *abstrait,* S.E.V.P.E.N., Paris, 1957.

Delco A., *La natura di Merleau-Ponty. A proposito di alcuni "inediti "del filosofo francese*, "Filosofia", n. 46, 3, 1995, pp. 185-212.

Delivoyatzis S., *Engagement ou neantisation? [Merleau-Ponty – Sartre*], "Philosophia" (Ath.), n. 17-18, 1987-88, pp. 161-181.

Delivoyatzis S., *La dialectique du phénomène. Sur Merleau-Ponty,* Meridiens Klincksieck, Paris 1987.

Delogu A., *Né rivolta né rassegnazione. Saggio su Merleau-Ponty*, ETS, Pisa, 1980.

Depraz N., *Chair de l'esprit de la chair chez Hegel, Schelling et Husserl*, "Revue Philosophique de Louvain", n. 94, 1, 1996, pp. 19-42.

Derossi, G., *Dalla percezione alla visione. L'ontologia negativa dell'ultimo Merleau-Ponty*, "Filosofia", 1964.

Di Fazio C., *The free body. Notes on Maurice Merleau-Ponty's phenomenology of moviment,* in "Chiasmi International", n. 17, 2015, pp. 327-345.

Di Martino C., *The body of freedom,* in "Chiasmi International", n. 11, 2009, pp. 419-430.

Di Rienzo C., *Il pensiero del corpo. Un percorso estetico nell'ultimo Merleau-Ponty,* in "Chiasmi International", n. 12, 2010, pp. 241-258.

Dillon M.C., *Merleau-Ponty and the Psychogenesis of the Self* "The Journal of Phenomenological Psycology", n. 9, 1978, pp. 84-98.

Dillon M.C., *Merleau-Ponty on Existential Sexuality. A critique*, "The Journal of Phenomenological Psychology", n. 11, 1, 1980, pp. 67-81.

Dillon M.C., *"Eye and Mind". The Intertwining of Vision and Thought*, "Man World "13, 1980, pp. 155-71.

Dillon M.C., *Merleau-Ponty and the Reversibility Thesis*, "Man World", 16, 1983, pp. 365-388.

Dillon M.C., *Merleau-Ponty and the Trascendence of Immanence. Overcoming the Ontology of Consciousness*, "Man World", 19, 1986, pp.395-412.

Dillon M.C., *Apriority in Kant and Merleau-Ponty*, "Kant Studien", n. 78, 1987, pp. 403-423.

Dillon M.C., *Merleau-Ponty's Ontology (Studies in Phenomenology and Existential Philosophy),* Indiana University Press, Bloomington 1988.

Dillon M.C., *Écart and Difference. Merleau-Ponty and Derrida on Seeing and Writing*, Humanities Press, New York 1996.

Dodeman C., *"Un selbst qui est unn autre": la figure d'Ulysse chez Paul Ricœur et Maurice Merleau-Ponty,* in "Chiasmi International", n. 17, 2015, pp. 257-274.

Dodeman C., *Une "philosophie militante" est-elle encore possible? Merleau-Ponty et l'échec des philosophies de l'histoire,* in "Chiasmi International", n. 19, 2018, pp. 39-56.

Dodeman, C., *La philosophie militante de Merleau-Ponty*, OUSIA, Bruxelles 2022, p. 161.

Duchene M., *Du corps objet au Corps expressif: l'ontologie phénoméologique de Merleau-Ponty*, in B. Feltz, D. Lambert (a cura di), *Entre le corps et l'esprit. Approche interdisciplinaire du Mind Body Problem,* Pierre Mardaga, Liège, 1994, pp. 33-52.

Dufourcq A., *"Sous les masques il n'y a pas de visage": l'éthique merleau-pontyenne entre probléme de l'altérité radicale, foi et institution,* in "Chiasmi International", n. 17, 2015, pp. 347-369.

Dufrenne M., *La peinture à l'oeuvre et l'enigme du corps*, "Revue d'Esthétique", 3-4, 1980, pp. 355-359.

Dufrenne M., *Comment voir l'histoire?* "Esprit" 6, 1982, pp. 45-52.

Duhan L., *Ambiguity of Time, Self, and Philosophical Explanation in Merleau-Ponty*, "Auslegung", 13, 1987, pp. 126-138.

Duportail G.F., *La sainte victoire de Cézanne,* in "Chiasmi International", n. 17, 2015, pp. 225-236.

Embree L., *The impression of causality: Merleau-Ponty on Michotte,* in "Chiasmi International", n. 11, 2009, pp. 311-320.

Enegren A., *Hanna Arendt lectrice de Merleau-Ponty* "Esprit", n. 6, 1982, pp. 154-155.

Erbetta A., *Educazione ed esistenza*, Il Segnalibro, Torino, 1998.

Erbetta A., *Il tempo della giovinezza. Situazione pedagogica e autenticità esistenziale,* La Nuova Italia, Milano-Firenze, 2001.

Escher Di Stefano A., *Il radicalismo filosofico come esperienza del vissuto in Merleau-Ponty*, Loffredo, Napoli 1981.

Escoubas E., *Merleau-Ponty: Die Kunst und das Phänomen [Husserl Merleau-Ponty]*, "Phänomenologische Forschung" n. 30, 1996, pp. 222-234.

Escoubas E., *Merleau-Ponty et Maldiney: deux figures de la phénoménologie,* in "Chiasmi International", n. 10, 2008, pp. 291-305.

Eslin J.C., *Critique de l'humanisme vertueux*, "Esprit", 6, 1982, pp. 7-20.

Esposito R., 1998. *Communitas. Origine e destino della comunità*, Einaudi, Torino 1998.

Esposito, R., *Per un pensiero istituente*, in "Discipline filosofiche", 2, 2019, pp. 9–30.

Esposito, R., *Pensiero istituente. Tre paradigmi di ontologia politica*, Einaudi, Turin 2020.

Esposito, R., *Istituzione*, il Mulino, Bologna 2021.

Ewald. F., (a cura di), *Sartre, Merleau-Ponty: les lettres d'une rupture*, "Magazine Litteraire", 320, 1994, p. 69.

Fergnani F., *Marxismo e filosofia dell'ambiguità*, "Il pensiero critico", 4, 1960 pp. 16-65.
Fergnani F., *Esistenza e corporeità nel pensiero di Maurice Merleau-Ponty*, in Id., (eds.), *Il corpo vissuto*, Il Saggiatore, Milano 1979, pp. 7-75.
Ferraris, A., *L'apertura all'alterità nella filosofia di Merleau-Ponty*, "Atti dell'Accademia delle Scienze di Torino", 1962-63.
Fink E., *Les concepts opératoires dans la phénoménologie de Husserl*, La Haye 1959.
Firenze A., *La filosofia dell'animalità in Heidegger e Merleau-Ponty,* in "Chiasmi International", n. 12, 2010, pp. 311-332.
Flores D'arcais P., *Prefazione* a M. Merleau-Ponty, *Umanismo e terrore*, Sugarco, Milano 1978, pp. 7-18.
Florival G., *L'uomo e la corporeità in Merleau-Ponty*, in Aa.Vv, *Il corpo, perché? Saggi sulla struttura corporea della persona*, XXXIII Congress by Centro di Studi Filosofici di Gallarate, 30 marzo-1aprile 1978, Morcelliana, Brescia 1979, pp. 237-254.
Florival G., *La philosophie de l'existence et du corps cbez Merleau-Ponty*, in P. Jonckheere (a cura di), *Phénoménologie et analyse existentielle*, Presses Universitaires de Louvain, Cabay, 1986, pp. 95-108.
Flynn B.C., *The Question of an Ontology of the Political: Arendt, Merleau-Ponty, Lefort*, "International Studies in Philosophy, n. 16, 1, 1984, pp. 1-24.
Fortuzzi M., *Istituire il futuro. Politica, istituzione e simbolo nel pensiero di Merleau-Ponty*, Mimesis, Milano 2023.
Foti V.M., *Painting and the Re-Orientation of Philosophical Thought in Merleau-Ponty*, "Philosophy Today", n. 24, 1980, pp. 114-120.
Foti V.M., *On Truth/Untruth in Heidegger and Merleau-Ponty*, "Research in Phenomenology, n. 13, 1983, pp. 185-198.
Foti V.M., *Heidegger's and Merleau-Ponty's Turn from Technicity to Arts* "Philosophy Today", 30, 1986, pp. 306-316.
Foti V.M., *Espression, alterity, and the phylosophy of nature in Merleau-Ponty's dialogue with the rationalists,* in "Chiasmi International", n. 11, 2009, pp. 279-291.
Franzini E., *La pittura e la differenza*, "Chiasmi International", n. 1, 1999, pp. 187-98.
Freud S., *Gesammelte Werke*, S. Fisher Verlag, Frankfurt, Imago Publishing Co, London 1948.
Galimberti U., *Il corpo*, Feltrinelli, Milano 1994.
Gambazzi P., *L'occhio e il suo inconscio*, Cortina, Milano 1977.
Gambazzi P., *Fenomenologia e psicoanalisi nell'ultimo Merleau-*

Ponty, "aut-aut", n. 232-233, 1989, pp. 105-129.

Gambazzi P., *La piega e il pensiero. Sull'ontologia di Merleau-Ponty*, "aut-aut", 262-263, 1994, pp. 21-47.

Gambazzi P., *Fissione dell'essere, essenze e visibilità assoluta. Individuazione ed ecceità della cosa nell'ultimo Merleau-Ponty*, in "Chiasmi International", n. 1, 1999, pp. 253-273.

Garelli J., *Il y a le monde*, "Esprit" n. 6, pp. 113-123, 1982.

Garelli J., *Inconscient phénoménologique et images surrealistes*, "Melusine" n. 11, 1990, pp. 175-183.

Garelli J., *Le lieu d'un questionnement. Appendice sur les rapports entre le precategoriel kantien et le preréflexif merleau-pontien*, "Les Cahiers de Philosophie", n. 7, 1989, pp, 107-140 e 141-143.

Garelli J., *Remarques sur mon rapport à l'œuvre de Merleau-Ponty*, in "Chiasmi International", n. 10, 2008, pp. 197-205.

Gargani, A., *Crisi della ragione*, Einaudi, Turin 1976, p. 33.

Gault S., *Tracing Merleau-Ponty's passage to ontology: a genetic investigation of* fundierung and stiftung, in "Chiasmi International", n. 19, 2018, pp. 345-369.

Geraets T.F., L'idea del "corpo proprio "e la struttura della filosofia secondo Merleau-Ponty, "Quaderni di Filosofia", n. 9, 10-11, 1984, pp. 63-72.

Geraets T.F., *Le retour à l'experience perceptive et le sens du primat de la perception*, "Dialogue", n. 15, 1976, pp. 595-607.

Geraets T.F., *Sens perçu, profondeur et réalité dans la "Phénoménologie de la perception* ", "Studi Filosofici", n. 4, 1981, pp. 231-247.

Geraets T.F., *Merleau-Ponty's Conception of Nature*, "Analecta Husserliana", n. 16, 1983, pp. 301-312.

Giovannangeli D., *Merleau-Ponty: la peinture et l'enigme du corps*, "Revue d'Esthétique", 21, 1992, pp. 211-215.

Gléonec, A., *L'institution et la passivité. Lectures de Merleau-Ponty*, Millon, Grenoble 2017.

Godani P., *Variazioni sul sorvolo. Ruyer, Merleau-Ponty, Deleuze e lo stato della forma,* in "Chiasmi International", n. 13, 2011, pp. 349-359.

Gregori I., *Merleau-Ponty's Phänomenologie der Sprache*, Karl Winter, Heidelberg 1977.

Gueroult M., (eds.), *Autopresentazione*, "aut-aut"1989, n. 232-33, pp. 5-12 (ed. or. *Un inédit de Merleau-Ponty*, "Revue de Métaphysique et de morale", pp. 1962, 401-409).

Gurvitch G., *Les tendances actuelles de la philosophie allemande*, Vrin, Paris 1930.

Halda B., *Thematique phénoménologique et implications. Husserl, Edith Stein, Merleau-Ponty*, Nauwelaerts, Bruxelles 1976.
Hartmann N., *Hegel et le problème de la dialectique du réel*, "Revue de Métaphysique et de morale" juillet-septembre 1931.
Hegel G. W. F., *Phenomenology of Spirit*, Transl. by A. V. Miller, Oxford University Press, Oxford 1977.
Heidegger M., *Sein und Zeit*, © Max Niemeyer Verlag, Tübingen 1927.
Heidegger M., *Wom Wesen des Grundes*, © Klosterman, Frankfurt am Main 1950.
Heidegger M., *Holzwege*, © Klosterman, Frankfurt am Main 1950.
Heidegger M.,*Vorträge und Aufsätze*, © Verlag Günther Neske, Pfulingen 1957.
Heidsieck F., *L'ontologie de Merleau-Ponty*, PUF, Paris 1971.
Hirose K., *La genèse du social et la donation charnelle du non-présentable. La portée de la notion d'institution chez Merleau-Ponty,* in "Chiasmi International", n. 12, 2010, pp. 333-345.
Hirose K., *Instituer le chiasme: à partir du cours sur Hegel de Maurice Merleau-Ponty,* in "Chiasmi International", n. 16, 2014, pp. 221-237.
Husserl, E., *Manoscritto K III*, 1, it. transl. by E. Filippini, Bompiani, Milan 1960.
Husserl, E., *The Crisis of European Sciences and Transcendental Phenomenology: An Introduction to Phenomenological Philosophy*; Translated, with an Introduction, by David Carr. Translated by David Carr, Northwestern University Press, Evanston, Ill. 1970.
Husserl, E., *Ideas Pertaining to a Pure Phenomenology and to a Phenomenological Philosophy*, Kluwer Academic Pub, Dordrecht 1983.
Husserl E., *Per la fenomenologia della coscienza interna del tempo*, Franco Angeli, Milano 1985.
Hyppolite J., *L'existence dans la "Phénoménologie de Hegel"*, "Revue des Etudes germaniques", 2; poi in Id, *Figures de la pensée philosophique. Ecrits* (1931-1968), Pue, Paris 1946, pp. 92-103.
Hyppolite J., *L'évolution de la pensée de Merleau-Ponty,* in Id., *Figures de la pensée philosophique. Ecrits* (1931-1968), PUF, Paris 1971, pp. 705-730.
Hyppolite, J., *Genesis and Structure of Hegel's Phenomenology of Spirit*, Transl. by S. Cherniak and J. Heckman, Northwestern University Press, Evanston, Illinois 1974.

Invitto G., *Merleau-Ponty politico: l'eresia programmatica*, Lacaita, Manduria 1971.
Invitto G., *Merleau-Ponty e la filosofa come vigilanza*, Milella, Lecce 1981.
Invitto G., *Le nozioni antisartriane di "occhio" e "visione" nell'ultimo Merleau-Ponty*, "Segni e comprensione", 1, 1987, pp. 47-50.
Invitto G., *Merleau-Ponty par lui-même. Un filosofo parla di sé*, in "Chiasmi International", n. 10, 2008, pp. 59-83.
Iori V., *Lo Spazio vissuto. Luoghi educativi e soggettività,* La Nuova Italia, Firenze 1996.
Jeanson, F., *La fenomenologia*, Sugar, Milano 1962.
Johnson G.A., *Merleau-Ponty's Early Aestetics of Historical Being: the Case of Cézanne*, "Research in Phenomenology", 17, 1987, pp. 211-225.
Johnson G.A., *The Colors of Fire: Depth and Desire in Merleau-Ponty's "Eye and Mind*", "The Journal of the British Society for Phenomenology", n. 25, 1, 1994, pp. 53-63.
Kakuni T., *Le corps aux limites de la raprésentation. Théorie du corps et de la peinture chez Merleau-Ponty,* in "Chiasmi International", n. 12, 2010, pp. 203-214.
Kirchmayr, R., (eds.) Sartre, J.-P., *Merleau-Ponty*, Cortina, Milano, 1999.
Kirchmayr, R., *Alle radici di un equivoco. Filosofia e politica in Sartre e Merleau-Ponty*, in "aut aut", 381, 2019, pp. 91–119.
Klee P., *Quaderno di schizzi pedagogici*, Vallecchi, 1925, Firenze, 1979.
Klee, P., *Pedagogical Sketchbook*. London, Faber And Faber, 2000.
Kockelmans J.J., *On the Function of Psychology in Merleau-Ponty's Early Works*, "Review of Existential Psychology & Psychiatry", n. 18, 1982-83, pp. 119-142.
Koestler, A., *Darkness at Noon*, MacMillan, London 1940.
Koestler A., *The Yogi and the Commissar*, MacMillan, London 1945.
Koffka K., *Principi di psicologia della forma*, Boringhieri, 1962, Torino, 1970.
Köhler W., *La psicologia della Gestalt*, Feltrinelli, 1929, Milano, 1961.
Kojève A., *La dialettica e l'idea della morte in Hegel*, Einaudi, Torino, 1947.
Kojève A., *Introduction to the Reading of Hegel: Lectures on the Phenomenology of Spirit*, transl. by James H. Nichols Jr., Cornell

University Press, N.Y. 1980.
Kristensen S., *Corps et symbolisation. La phylosophie du dernier Merleau-Ponty et la question d'une épistémologie de la chair,* in "Chiasmi International", n. 11, 2009, pp. 321-337.
Kristensen S., *L'incoscient machinique et l'idée d'une ontologie politique de la chair,* in "Chiasmi International", n. 16, 2014, pp. 127-147.
Kruks S., *Merleau-Ponty, Hegel and the Dialectic*, "The Journal of the British Society for Phenomenology", 7, 1976, pp. 96-110.
Kruks S., *Merleau-Ponty: a Phenomenological Critique of Liberalism*, "Philosophy and Phenomenological Research", n. 37, 1977, pp. 394-407.
Kruks S., *The Political Philosophy of Merleau-Ponty*, Humanities Press, Atlantic Highlands, 1981.
Kruks S., *A Study of the Political Philosophy of Merleau-Ponty*, Garland, New York 1987.
Kruks S., *Marcel and Merleau-Ponty. Incarnation, Situation and the Problem of History*, "Human Studies", n. 10, 1987, pp. 225-245.
Kruks S., *Simone de Beauvoir entre Sartre et Merleau-Ponty*, "Les Temps Modernes", n. 520, 1989, pp. 81-103.
Kruks S., *Situation and Human Existence, Freedom, Subjectivity and Society*, Unwin Hyman, London 1990.
Kwant R. C., *We Inhabit the World*, "Humanitas", 12, 1976, pp. 299-309.
Kwant, R. C., F*rom phenomenology to metaphysics. An inquiry into the last period of Merleau-Ponty's philosophical life*, Duquesne University Press, Pittsburgh 1966.
Lacan, J., *Seminar book XI,* transl. by A. Sheridan, ed. by J.A. Miller. W.W. Norton & Company, New York 1981.
Landes D.A., *Le sujet de la sensation et le sujet résonant. Communion et renvoi chez Merleau-Ponty et Nancy,* in "Chiasmi International", n. 19, 2018, pp. 143-161.
Lanfredini R., *Divenire di Merleau-Ponty. Filosofia di un soggetto incarnato,* Guerini e associati, Milano 2011.
Lanfredini R. and Vitale S., (eds.), *Pensare in ultimo. Merleau-Ponty dopo Merleau-Ponty*, Moretti & Vitali, 2014.
Lanfredini R., *Emotion and affection between phenomenology and psychoanalysis: behavior, body, memory,* in "Chiasmi International", n. 18, 2016, pp. 33-48.
Lapierre C., *Entre conscience et inconscient. Une lecture du problème de la passivité chez Sartre et Merleau-Ponty,* in "Chiasmi International", n. 17, 2015, pp. 95-132.

Larison M., *Autor du concept de nature dans le dernier Merleau-Ponty,* in "Chiasmi International", n. 7, 2005, pp. 393-405.

Larison M., *L'Être en forme. Dialectique et phénoménologie dans la derneère philosophie de Merleau-Ponty*, Editions Mimesis, Milan 2016.

Larison M., *Stiftung et pensée du social: a propos de phénoménologie merleau-pontienne de l'istitution,* in "Chiasmi International", n. 18, 2016, pp. 363-376.

Lawlor L., *The End of Ontology Interrogation in Merleau-Ponty et Deleuze* "Chiasmi International", 1, 1999, pp. 233-252.

Le Goff J.P., *Le paradoxe du langage et de l'étre brut. Question sur une ontologie*, "Les Cahiers de Philosophie", 7, 1989, pp. 69-84.

Lefebvre H., *Il materialismo dialettico*, Einaudi, Torino 1949.

Lefebvre, H., *Dialectical Materialism*, University of Minnesota Press, Minneapolis 1968.

Lefeuvre, M., *Merleau-Ponty au delà de la phénoménologie du corps, de l'être et du langage,* Université de Lille III, Lille 1977.

Lefort C., *L'idée d'"ètre brut "et d'"esprit sauvage*", "Les Temps Modernes "184-185, 1961, pp. 225-87.

Lefort C., *Sur une colonne absente. Ecrits autour de Merleau-Ponty*, Gallimard, Parigi. 1978.

Lefort C., *D'un doute à l'autre*, "Esprit", 6, 1982, pp. 23-30.

Lefort C., *Philosophie et non-philosophie*, "Esprit, 6, 1982, pp. 101-112.

Lefort C., *Merleau-Ponty en accusation. Mise au point*, "Esprit", 10, 1985, pp. 83-86.

Lefort C., *Introduzione*, M. Merleau-Ponty, *Il visibile e l'invisibile*, Bompiani, Milano 1993.

Leland D., *Husserl, Heidegger, Sartre, Merleau-Ponty. On Phenomenology and the Problem of Intentionality,* Hackett, Indianapolis 1989.

Leoni F., *L'esperienza della filosofia. Senso e significato della parola filosofica nell'ultimo Merleau-Ponty,* in "Chiasmi International", n. 10, 2008, pp.183-193.

Leoni F., *La machine et la vie. Histoire du mouvement et histoire de la métaphysique,* in "Chiasmi International", n. 11, 2009, pp. 293-310.

Leoni F., *Introduction: un autre inconscient,* in "Chiasmi International", n. 18, 2016, pp. 21-29.

Leoni F., *Psychoanalysis as experimental ontology,* in "Chiasmi International", n. 18, 2016, pp. 105-113.

Letellier I., *Le réel et la non-altérité de l'autre. Lacan avec Merleau-*

Ponty, in "Chiasmi International", n. 17, 2015, pp. 133-146.

Levin D.M., *Eros and psyche. A Reading of Merleau-Ponty*, "Review of Existential Psychology & Psychiatry", 18, 1982-83, pp. 219-239.

Levin D.M., *Sanity and Mith in Affective Space. A Discussion of Merleau- Ponty,* "Philosophical Forum", n. 14, 1982-83, pp. 157-189.

Levinas E., *De l'intersubjectivité. Notes sur Merleau-Ponty,* in Id., *Hors Sujet*, Fata Morgana, Montpellier, 1987-89, pp. 143-153.

Lingis A., *Phenomenological explanations*, M. Nijhoff, L'Aia 1986.

Lisciani-Petrini E., *Merleau-Ponty-Bergson: un dialogo "se faisant*", "Il Pensiero", XXXIII, 1993, pp. 67-93.

Lisciani-Petrini E., *L'invisibile di Cézanne*, "Il Pensiero", XXXVI, 1, 1997, pp. 23-34.

Lisciani-Petrini, E., *Presentazione de Il carteggio della rottura*, "MicroMega", 1, 1997, pp. 193–198.

Lisciani-Petrini E., *Attività/Passività: l'invisibile di Merleau-Ponty,* in "Chiasmi International", n. 1, 1999, pp. 169-185.

Lisciani-Petrini E., *La passione impolitica della politica, Merleau-Ponty tra "filosofia e non-filosofia*", in VV.AA., *Nichilismo e politica*, Laterza, Roma-Bari, 2000, pp. 55-73.

Lisciani-Petrini E., *La passione del mondo*. Saggio su Merleau-Ponty, ESI, Naples 2002.

Lisciani-Petrini, E., *Risonanze. Ascolto, Corpo, Mondo*, Mimesis, Milan 2006.

Lisciani-Petrini E., *Per un "Dérèglement de tous les sens". Merleau-Ponty: le sinestesie e l'impersonale,* in "Chiasmi International", n. 10, 2008, pp. 109-127.

Lisciani-Petrini E., *"La voix de personne". Au-delà des frontières du sujet,* in "Chiasmi International", n. 11, 2009, pp. 348-358.

Lisciani-Petrini, E., *Fuori della persona*. L'impersonale in Bergson, Merleau-Ponty, Deleuze, "Daimon", 44, 2012, pp. 73–88.

Lisciani-Petrini, E., *Hacia el sujeto impersonal*, "Pléyade", vol. 9, 2012, pp. 137–159.

Lisciani-Petrini, E., *Verso il soggetto impersonale*, "Filosofia politica", vol. 1, 2012, pp. 39–50.

Lisciani-Petrini, E., *Vita quotidiana. Dall'esperienza artistica al pensiero in atto*, Bollati Boringhieri, Turin 2015.

Lisciani-Petrini, E., *Sartre, Merleau-Ponty: le lettere della rottura*, in " aut aut", 381, 2019.

Lisciani-Petrini, E., *Potenza dell'istituzione,* "Discipline filosofiche", 2, 2019, pp. 71–99.

Lisciani-Petrini, E., *Merleau-Ponty/Sartre: una insanabile divergenza filosofico-politica*, "aut aut", 381, 2019, pp. 61–90.

Lukacs, G., *History and Class Consciousness*, Merlin Press, 1967.

Luquet G.H., *Il disegno infantile*, Armando, Roma 1969.

Madison G.B., *La phénoménologie de Merleau-Ponty. Une recherche des limites de la conscience,* Klincksieck, Paris, 1973.

Madison G.B., *Merleau-Ponty et la contre-tradiction*, "Dialogue", 17, 1978, pp. 456- 419. Madison G.B., *Du. corps à la chair: Maurice Merleau-Ponty*, "Analecta Husserliana", n. 21, 1986, pp. 167-88.

Madison G.B., *Merleau-Ponty et la deconstruction du logocentrisme*, "Laval Thélogique et Philosophique", n. 46, 1, 1990, pp. 65-79.

Malabou C., *Phantom Limbs and Plasticity: Merleau Ponty and current neurobiology,* in "Chiasmi International", n. 17, 2015, pp. 41-52.

Mancini, S. *L'ontologia indiretta dell'ultimo Merleau-Ponty*, in VV.AA., *Merleau-Ponty: Esistenza, filosofia, politica*, ed. by Invitto, G. Guida, Naples 1982.

Mancini S., *Il narcisismo della carne, Merleau-Ponty e l'interpretazione ontologica della pittura*, "aut aut", n. 202-203, 1984, pp. 136-148.

Mancini S., *L'assoluto dialettico in Merleau-Ponty,* "Il Centauro", n. 10, 1984, pp. 90-110.

Mancini S., *Sempre di nuovo. Merleau-Ponty e la dialettica dell'espressione*, Franco Angeli, Milano 1987.

Marinotti A, *Percezione e conoscenza. Linee della fenomenologia genetica di M. Merleau-Ponty*, Alfani, Firenze 1983.

Martone A., *Verità e comunità in Maurice Merleau-Ponty*, La Città del Sole, Napoli 1998.

Marx K., *Scritti giovanili (Per la critica della filosofia del diritto di Hegel; Manoscritti economico-filosofici del 1844; Prima tesi su Feuerbach*), cura di A. Del Noce, G.A, Riestra, Japadre, L'Aquila 1975.

Matsuba S., *Possibilité de la "communauté charnelle",* in "Chiasmi International", n. 11, 2009, pp. 401-417.

Mc Weeny J., *Introduction. The body of our time,* in "Chiasmi International", n. 18, 2016, pp. 137-149.

Mercury J.-Y., *L'expressivité chez Merleau-Ponty. Du corps à la peinture*, L' Harmattan, Paris 2000.

Merleau-Ponty, M., *Humanism and Terror: An Essay on the Communist Problem*, Beacon Press, Boston 1947.

Merleau-Ponty, M., *Communisme et anti-communisme*, "Les Temps

Modernes", 34, 1948, pp. 175-188.
Merleau-Ponty, M., (with J.-P. Sartre), *Les jours de notre vie*, "Les Temps Modernes", 51, 1950, pp. 1153-1168.
Merleau-Ponty, M., *Phenomenology of Perception,* transl. by Colin Smith, Routledge, London and New York 1962.
Merleau-Ponty, M., *Un inedit,* in "Revue de Metaphisique et de Morale" n. 4, 1962.
Merleau-Ponty, M., *Signs,* Northwestern University Press, Evanston, Ill., 1964.
Merleau-Ponty, M., *Eye and Mind*, in J. E. Edie (Ed.), *The Primacy of Perception*, Northwestern UP Evanston, Evanston, 1964.
Merleau-Ponty, Maurice. *The Visible and the Invisible*, Northwestern University Press, Evanston, Ill., 1968.
Merleau-Ponty, M., *Themes from the Lectures at the Collège de France, 1952-1960*, Northwestern University Press, Evanston, Ill. 1970.
Merleau-Ponty, M., *Adventures of the Dialectic*. Northwestern University Press, Evanston, Ill., 1973.
Merleau-Ponty, M., *In Praise of Philosophy and Other Essays*, Northwestern University Press, Evanston, Ill., 1988.
Merleau-Ponty, M., *Philosophy and Non-Philosophy since Hegel*, in Silverman, H. J., *Philosophy and Non-Philosophy since Merleau-Ponty*, Routledge, New York and London 1988.
Merleau-Ponty, M., *The Prose of the World*. Evanston, Northwestern University Press, 1991.
Merleau-Ponty, M., *Sense and Non-Sense*, Northwestern University Press, Evanston, Ill. 1992.
Merleau-Ponty, M., *Institution and Passivity*. Northwestern University Press, 2010.
Merleau-Ponty, M., *The Structure of Behavior*. Pittsburgh, Duquesne University Press, 2015.
Merleau-Ponty, M., *The Sensible World and the World of Expression: Course Notes from the Collège de France, 1953*, transl. by B. A. Smyth, Northwestern University Press, Evanston, Ill. 2020.
Messori R., *In punta di parole. Figura e metafora in Maurice Merleau-Ponty e Paul Ricœur,* in "Chiasmi International", n. 17, 2015, pp. 239-255.
Miano F., *L'etica della situazione come etica della responsabilità in Merleau-Ponty*, "Studium", n. 80, 6, 1984, pp. 863-876.
Michalet J., *La chair comme "plissement du dehors". La lecture deleuzienne du dernier Merleau-Ponty,* in "Chiasmi International", n. 13, 2011, pp. 241-258.

Michotte A., *The Perception of Causality*, Routledge, London 2017.
Minkowski E., *Il tempo vissuto. Fenomenologia e psicopatologia*, Einaudi, 1933, Torino 1970.
Monferrand, F., *Politiser l'expérience. Merleau-Ponty, Socialisme ou barbarie et l'expérience prolétarienne*, in "Chiasmi International", 19, 2018, pp. 87–100.
Mongin O., *Depuis Lascaux*, "Esprit", 6, 1982, pp. 67-76.
Mongin O., *Harmoniques*, "Esprit", 6, 1982, pp. 146-153.
Moravia S., *La crisi della generazione sartriana*, "Rivista di Filosofa", 58, 1967, pp. 426-470.
Moravia S., *Introduzione* a Freud, S. *Filosofia e Psicoanalisi*, edited by S. Moravia (it. transl.), La Nuova Italia, Firenze 1981.
Moreau J., *La presence du monde et l'ouverture a L'Etre,* "Les Etudes Philosophiques", 1959.
Morin M.E., *Corps propre or corpus corporium: unity and dislocation in the theoris of embodiment of Merleau-Ponty and Jean-Luc Nancy,* in "Chiasmi International", n. 18, 2016, pp. 333-351.
Morris D., *The Fold and the body schema in Merleau-Ponty and Dynamic Systems Theory,* in "Chiasmi International", n. 1, 1999, pp. 275-286.
Morris D., *What is living and what is non-living in Merleau-Ponty's philosophy of movement and expression,* in "Chiasmi International", n. 7, 2005, pp. 225-239.
Morris D., *Ecstatic body, ecstatic nature: perception as breaking with the world,* in "Chiasmi International", n. 8, 2006, pp. 201-217.
Morris D., *The chirality of being: exploring a Merleau-Pontyan ontology of sense,* in "Chiasmi International", n. 12, 2010, pp. 165-182.
Morris D., *Bringing phenomenology down to earth: passivity, development, and Merleau-Ponty's transformation of phylosophy,* in "Chiasmi International", n. 16, 2014, pp. 25-38.
Nepi P., *Merleau-Ponty tra il visibile e l'invisibile*, Studium, Roma 1984.
Neri G.D., *Storia e possibilità*, "aut-aut", n. 232-233, 1984, pp. 87-103.
Neri G.D., *Dialettica dell'espressione. Un saggio recente su Merleau-Ponty*, "aut-aut", 232-233, 1989, pp. 147-151.
O'connor D.T, *Ambiguity and the Search for Origins* "The Journal of the British Society for Phenomenology", n. 9, 1978, pp. 102-110.

O'connor D.T, *Merleau-Ponty and the Problem of the Unconscious*, "Research in Phenomenology", n. 10, 1980, pp. 77-88.

O'neill J., *Tbe Specular Body. Merleau-Ponty and Lacan on Infant Self and Other*, "Sinthese", n. 66, 2, 1986, pp. 201-17.

Olesen S.G., *Note sur la tàche de l'historien de la phylosophie,* in "Chiasmi International", n. 12, 2010, pp. 367-392.

Olkowski D., *Merleau-Ponty's Freudianism. From te Body of Consciousness to the Body of Flesh. Special Issue: Merleau-Ponty ans Psychology*, "Review of Existential Psychology & Psychiatry", n. 18, 1-3, 1982-83, pp 97-116.

Paci E., *Introduzione*, in M. Merleau-Ponty, *Elogio della filosofia*, Paravia, Torino, 1958, pp. V-XL.

Paci E., *Introduzione*, in M. Merleau-Ponty, *Senso e non senso*, Il Saggiatore, Milano, 1962.

Paci E., *Merleau-Ponty, Lukàcs e il problema della dialettica*, "aut aut", 66, 1961, pp. 498-516.

Paci E., *Ingens sylva*, Bompiani, Milano, 1994.

Pangallo M., *Il problema filosofico dell'alterità. Saggio sul solipsismo e l'intersoggettività in Maurice Merleau-Ponty*, La Goliardica, Roma1990.

Peillon V., *La tradition de l'Esprit. Itineraire de Maurice Merleau-Ponty*, Grasset, Paris 1994.

Petronella Foultier A., *Incarnated meaning and the notion of Gestalt in Merleau-Ponty's phenomenology,* in "Chiasmi International", n. 17, 2015, pp. 53-75.

Pfeiffer M.L., *Innocence, Guilt and Totalitarism* [M, Merleau-Ponty], "Phenomenological Inquiry", 16, 1992, pp. 77-82.

Philippe, M.-D. *Une philosophie de l'Être est-elle encore possible?*, Tequi, Paris 1975.

Pierron J.P., *La prose du monde ou le monde comme un texte? Ricœur, lecteur de Merleau-Ponty,* in "Chiasmi International", n. 17, 2015, pp. 309-323.

Pombo F., *Cézanne: a proposito de l'oeil et l'esprit*, "Coloquio Artes", 93, 1992, pp. 38-43 e 80-82.

Pompeo Faracovi, O., *Eresia e riformismo in Merleau-Ponty*, "Il Ponte", 9, 1972, pp. 1235–1259.

Pompeo Faracovi, O., *Filosofia e politica*, in VV.AA., *Esistenza, filosofia, politica*, edited by G. Invitto, Guida, Naples 1982, pp. 173–197.

Pontremoli E., *L'excès du visible. Une approche phénoménologique de la photogénie*, Millon, Grenoble, 1996.

Popper, K., *The Logic of Scientific Discovery*, Routledge, London-

New York 2002.

Potempski J., *Art and the sense of being from Merleau-Ponty to Jean-Luc Nancy,* in "Chiasmi International", n. 19, 2018, pp. 163-181.

Rabil A., *Merleau-Ponty, Existentialist of the Social World*, Columbia University Press, New York 1967.

Ramirez M.T., *El quiasmo: ensayo sobre la filosofia de Maurice Merleau-Ponty*, Morelia, Michoacan 1994.

Ramirez M.T., *La "veridad "de la fenomenologia* (Merleau-Ponty y Husserl), "Iztapalapa", 14,33, 1994, pp. 51-62.

Ramirez M.T., *La fenomenologia de la percepción de Merleau-Ponty como estética*, "Revista de Filosofia Mexicana", n. 28, 84, 1995, pp. 332-362.

Ramirez M.T., *Scienza e carnalità,* in "Chiasmi International", n. 8, 2006, pp. 43-60.

Reineke M., *Lacan, Merleau-Ponty and Irigaray. Reflections on a Specular Drama*, "Auslegung", 14, 1987, pp. 67-85.

Ret D., *La phénoménologie existentielle de Merleau-Ponty, L'ambiguité de l'existence humaine, c'est aussi l'ambiguité de la liberté*, "Freiburger Zeitschrift für Philosophie und Theologie", 35, 1988, pp. 137-157.

Revault D'allonnes M., *Révolution et histoire chez Merleau-Ponty et Hanna Arendt*, "Les Cahiers de Philosophie", 4, pp. 1987, 147-160.

Revault D'allonnes M., *Le doute de Merleau-Ponty*, "Les Temps Modernes", 531-533, (Témoins de Sartre), I, 1990, pp. 551-68.

Revel, J., *Un registre ouverte*, in "Esprit", 319, 2023, pp. 77–89.

Richir M., *Le sens de la phénoménologie dans "Le visible et l'invisible*", "Esprit ", 6, 1982, pp. 124-145.

Richir M., Merleau-Ponty: *un tout nouveau rapport la psycanalyse*, "Les Cahiers de Philosophie", 7, 1989, pp. 155-187.

Ricoeur P., *Par-délà Husserl et Heidegger,* "Les Cahiers de Philosophie", 7, 1989, pp. 17-23.

Riley M.D., *The Truth of the Body. Merleau-Ponty on Perception, Language and Litterature*, "*Analecta Husserliana*", 18, 1984, pp. 479-493.

Robinet, A., *Che cosa ha veramente detto Merleau-Ponty*, it. transl., Ubaldini Editore, Rome 1973.

Rocca E., *L'Essere e il giallo, Intorno a Merleau-Ponty*, Pratiche, Parma 1993.

Rosenthal S.B., Bourgeois P.L., *Peirce, Merleau-Ponty and Perceptual Experience. A Kantian Heritage*, "International

Studies in Philosophy", n. 19, 3, 1987, pp. 33-42.

Rosenthal S.B., Bourgeois P.L., *Mead and Merleau-Ponty. Towards a Common Vision*, State University of New York Press, Albany 1991.

Rosenthal S.B., Bourgeois P.L., *The World of Language. Merleau-Ponty and Mead*, "Phenomenological Inquiry", 1992, 16, pp. 96-110.

Rosenthal S.B., Bourgeois P.L., *The World of Truth. Merleau-Ponty and Mead*, "Southwest Philosophy Review", n. 10, 2, 1994, pp. 49-58.

Ross H., *Merleau-Ponty and J.P Sartre on the Nature of Consciousness*, "Cogito", n. 3, 4, 1985, pp. 115-121.

Rouger F., *La politique et l'histoire chez Maurice Merleau-Ponty*, "Les Études Philosophiques", n. 4, 1979, pp. 437-452.

Rouse J., *Merleau-Ponty and the Existential Conception of Science*, "Synthese", n. 66, 2, 1986, pp. 249-272.

Rovatti P.A., *L'occhio e lo sguardo*, in Id., *Abitare la distanza*, Feltrinelli, Milano 1994, pp. 54-66.

Salsa A., *L'ambiguità del corpo nelle analisi fenomenologiche di Maurice Merleau-Ponty*, in A. Dentone, M. Schiavone (eds.), *Segno, simbolo, sintomo, comunicazione. Implicanze e convergenze fra filosofia, psichiatria, psicoanalisi*, "Atti del Convegno", Chiavari, 5-7 dicembre 1984, Esagraph, Genova, 1985, pp. 317-322.

Salsa A., *The "Life World" as a Moral Problem in Merleau-Ponty*, "Analecta Husserliana", 22, 1987, pp. 131-137.

Sartre J. P., *Che cos'è la letteratura*? Il Saggiatore, Milano 1947.

Sartre J. P., *Critica della ragione dialettica*, Il Saggiatore, Milano 1990.

Sartre J. P., *Les communistes et la paix*, "Les Temps Modernes", 1952, pp. 695-763 e 1729-1819.

Sartre J. P., *Materialisme et révolution*, "Les Temps Modernes"1953, pp. 1537-1563, e 1-32.

Sartre, J.-P., *Merleau-Ponty vivant*, "Les Temps Modernes", 184-185, 1961, pp. 304-376.

Sartre J.P., *L'essere e il nulla*, Il Saggiatore, Milano, 1991.

Sartre J. P., *Merleau-Ponty vivo*, Cortina, Milano, 1999.

Sartre J.-P., – Merleau-Ponty, M., *Les lettres d'une rupture*, ed. by F. Ewald, in "Magazine Litteraire", 320, 1994.

Sartre J. P., *Le lettere della rottura*, "Micromega", 1, 1994, pp 199-221.

Saussure (de) F., *Cours de linguistique générale*, Payot, Paris 1916.

Schmidt J., *Maurice Merleau-Ponty. Politics, Phenomenology and* Ontology, "Human Studies", n. 6, 1983, pp. 295-308.

Schmidt J., *Maurice Merleau-Ponty between Phenomenology and Structuralism*, St. Martin's Press, New York 1985.

Schneider M., *Le visible et le tangible comme paradigmes du savoir* (M. Merleau-Ponty), "Analecta Husserliana", n. 20, 1986, pp. 297-311.

Scotti, G. *Originarietà e relazione in Merleau-Ponty*, "aut-aut", n° 39, 1957.

Scotti, G., *Sulla percezione in Merleau-Ponty*, "aut-aut", n. 41,1957.

Shällibaum, U., *La possibilité de la réflexion philosophique*, in VV.AA., *Merleau-Ponty. Le philosophe et son langage*, in "Recherche sur la philosophie et le langage", n. 15, 1993, pp. 339-356.

Sichère B., Merleau-Ponty *ou le corps de la philosophie,* Grasset, Paris 1982.

Sichère B., *Quelques remarques sur la pensée politique Merleau-Ponty*, "Cah. Philos. Pol. Jur", n. 2, 1982, pp. 181-194.

Silverman H.J., *Heidegger and Merleau-Ponty Interpreting Hegel*, "Research in Phenomenology", 7, 1977, pp. 209-224.

Silverman H.J., *Imagining Perceiving and Remembering*, "Humanitas", 14, 1978, pp. 197-207.

Silverman H.J., *Merleau-Ponty On Language and Communication* (1947-1948), "Research in Phenomenology", 9, 1979, pp. 168-181.

Silverman H.J., *Merleau-Ponty and the Interrogation of Language*, "Research in Philosophy", 10, 1980, pp. 122-141.

Silverman H.J., (a cura di), *The Horizons of Continental Philosophy. Essays on Husserl, Heidegger, and Merleau-Ponty*, Kluwer, Dordrecht 1988.

Silverman H.J., *Excessive responsability and the sense of the world (Merleau-Ponty and Nancy),* in "Chiasmi International", n. 10, 2008, pp. 307-319.

Silverman H.J., Barry J., (e.d.s.), *M. Merleau-Ponty. Text and Dialogues*, Humanities Press, Atlantic Highlands 1992.

Simms E.M., *The Infant's Experience of the World Stern, Merleau-Ponty and the Phenomenology of the Preverbal Self*, "Humanistic Psychologist", n. 21, 1, 1993, pp. 26-40.

Simms E.M., *Egocentric language and the upheaval of speech. A Merleau-Ponty inspired study of language acquisition,* in "Chiasmi International", n. 12, 2010, pp. 287-309.

Sini C., *Immagini di verità*, Spirali/Vel, Milano 1990.

Sini C., *Introduzione*, in M. Merleau-Ponty, *La prosa del mondo*, Editori Riuniti, Roma, 1984, pp. 7-16.
Sini, C., *Il silenzio e la parola*, Marietti, Genova 1989.
Sini C., *Filosofia e scrittura*, Laterza, Roma-Bari 1994.
Slatman J., *The sense of life: Husserl and Merleau-Ponty touching and being touched,* in "Chiasmi International", n. 7, 2005, pp. 305-325.
Sordini A., *Pittura e metafisica nell'ultimo Merleau-Ponty*, in *Il corpo vissuto*, F. Fergnani (eds.) Il Saggiatore, Milano 1979, pp. 195-201.
Sordini A., *I simboli e il corpo, Merleau-Ponty filosofo dell'espressione*, "Fenomenologia e Società", n. 7, 1985, pp. 67-86.
Spurling L., *Phenomenology and the Social World. The Philosophy of Merleau-Ponty and its Relation to the Social Sciences*, Routledge and Kegan Paul, London-Boston 1977.
Stawarska B., *Dialogue at the limit of phenomenoly,* in "Chiasmi International", n. 11, 2009, pp. 145-156.
Taddio, L., *Maurice Merleau-Ponty, Feltrinelli*, Milano 2024.
Tagore S., *The Eco of Silence. Toward a Reconstruction of Merleau-Ponty's Philosophy of History*, "International Philosophical Quarterly", n. 31, 4, 1991, pp. 427-434.
Taioli R., *L'Oriente e la filosofia in Merleau-Ponty*, "Segni e comprensione", X, 28, 1996, pp. 5-15.
Taminiaux J., *Merleau-Ponty de la dialectique à l'hyperdialectique*, in Id., *Recoupements*, Ousia, Bruxelles 1982, pp. 91-117.
Taminiaux J., *Was Merleau-Ponty on the way from Husserl to Heidegger?,* in "Chiasmi International", n. 11, 2009, pp. 21-31.
Thebaud J.L., *La chair et l'infini. J.F. Lyotard et Merleau-Ponty,* "Esprit", n. 6, 1982, pp. 158-162.
Thierry Y., *Du corps parlant. Le langage chez Merleau-Ponty,* Ousia, Bruxelles 1987.
Thierry Y., *Situation du philosophe (M. Merleau-Ponty)*, "Les Cahiers de Philosophie", n. 7, 1989, pp. 7-16.
Thomas-Fogiel I., *Merleau-Ponty: de la perspective au chiasme, la rigueur épistémique d'une analogie,* in "Chiasmi International", n. 13, 2011, pp. 381-406.
Tilliette, X., *Merleau-Ponty ou la mesure de l'homme,* Edition Seghers, Paris 1970.
Toadvine T., *The melody of life and the motif of philosophy,* in "Chiasmi International", n. 7, 2005, pp. 263-279.
Treguier J.M., *Le corps selon la chair Phénoménologie et ontologie*

chez Merleau-Ponty, Kimé, Paris 1996.

Trigg D., *The role of the heart in Merleau-Ponty's archaeological phenomenology,* in "Chiasmi International", n. 16, 2014, pp. 255-272.

Tuppini T., *Merleau-Ponty e Nancy sul senso,* in "Chiasmi International", n. 11, 2009, pp. 81-97.

Umbelino L.A., *La nature indirecte et les couleurs de l'invisible. Notes à propos du projet ontologique de Merleau-Ponty,* in "Chiasmi International", n. 12, 2010, pp. 273-285.

Vallier R., *Institution: the significance of Merleau-Ponty's 1954 course at the collège de France,* in "Chiasmi International", n. 7, 2005, pp. 281-303.

Vanzago L., *Process and events of nature. Merleau-Ponty and Deluze as readers of Whitehead,* in "Chiasmi International", n. 10, 2008, pp. 227-250.

Vanzago L., *Nature, negativity, event. The question of the emergence of the centre in the ontology of the flesh,* in "Chiasmi International", n. 11, 2009, pp. 171-183.

Vanzago L., *The many faces of movement. Phenomenological and ontological questions concerning the relation between Perception, Expression and movement in Merleau-Ponty's lecture course on the sensible world and the world of expression,* in "Chiasmi International", n. 12, 2010, pp. 111-126.

Vanzago L., *Merleau-Ponty*, Carocci, Roma 2012.

Vanzago L., *Raw being and the darkness of nature. Merleau-Ponty's appropriation of Schelling,* in "Chiasmi International", n. 16, 2014, pp. 239-251.

Vanzago L., *The flesh between. Some remarks on André Green's remarks on Merleau-Ponty' ontology and its relationship to psychoanalysis,* in "Chiasmi International", n. 18, 2016, pp. 49-61

Vanzago, L., *The Voice of No One. Merleau-Ponty on Nature and Time*, Mimesis International, London–Milan 2017.

Vanzago L., *Leggere Il visibile e l'invisibile di Merleau-Ponty*, Ibis, Como-Pavia 2020.

Varela Ch. R., *Harre and Merleau-Ponty. Beyond the Absent Moving Body in Embodied Social Theory*, "Journal for the Theory of Social Behaviour", n. 24, 2, 1994, pp. 167-185.

Venier V., *Archeologia della coscienza. Nota in margine alla fenomenologia di Merleau-Ponty,* in "Chiasmi International", n. 7, 2005, pp. 329-339.

Villani T., *Linguaggio e verità, Il chiasma: il sentire della carne,*

"Fenomenologia e Società", XVII, n. 3, 1994, pp. 48-59.
Vitiello V., *La voce riflessa. Logica ed etica della contraddizione*, Lanfranchi, Milano 1994.
Vitiello V., *Non dividere il si dal no, Tra filosofia e letteratura*, Laterza, Roma-Bari 1996.
Vuillerod J.B., *Merleau-Ponty hégélien?*, in "Chiasmi International", n. 19, 2018, pp. 101-114.
VV.AA., Filosofia, esistenza, politica, Guida, Napoli 1982.
VV.AA., Merleau-Ponty, *le physique et le corporel*, Aubier, Paris 1988.
VV.AA., "*Ontology and Alterity*" in Merleau-Ponty, Northwestern University Press, Evanston, 1990.
VV.AA., *La prosa del mondo*. Omaggio a Maurice Merleau-Ponty, Ed, Quattroventi, Urbino 1991.
VV.AA., *Merleau-Ponty vivant*, State University of New York Press, Albany 1991.
VV.AA., *Merleau-Ponty phénoménologie et expériences,* Millon, Grenoble 1992.
VV.AA., *Merleau-Ponty Hermeneutics and Postmodernism*, State University of New York Press, Albany1992.
VV.AA., *Negli specchi dell'Essere. Saggi sulla filosofia di Merleau-Ponty*, Hestia edizioni, Cernusco (Como), 1993.
VV.AA., *The Merleau-Ponty Aesthetics Reader. Philosophy and Painting*, Northwestern University Press, Evanston, 1993.
VV.AA., *Merleau-Ponty, le philosophe et son langage*, "Recherches sur la philosophie et le langage", 15, Grenoble 1993.
VV.AA., *Merleau-Ponty in Contemporary Perspective*, Kluwer Academic Publishers, Dordrecht, Boston, London 1993.
VV.AA., *Merleau-Ponty. Difference, Materiality*, Humanities Press, New Jersey 1996.
VV.AA., *Merleau-Ponty aux frontières de l'invisible*, Mimesis, Milan 2003.
Waelhens A., De, *Una filosofia dell'ambiguità*, in M. Merleau-Ponty, *La struttura del comportamento*, Bompiani, Milano 1963.
Waelhens A., De, *Note sur l'évolution de la notion d'inconscient chez Sartre et Merleau-Ponty, in Les philosophies de l'existence et les limites de l'homme*, Vrin, Paris 1981, pp. 73-87.
Wahl, J., *Le Malheur de la conscience dans la philosophie de Hegel*, PUF, Paris 1951.
Waldenfels, B., *Twovards an Open Dialectics*, in "Dialectics and Humanism", n. 3, 1976, pp. 91-101.
Waldenfels, B., *Vérité à faire. La questione della verità in Merleau-*

Ponty, in VV. AA., *La prosa del mondo. Omaggio a Merleau-Ponty,* Quattroventi, Urbino 1991.

Waldenfels B., *Faire voir par les mots. Merleau-Ponty et le tournant linguistique,* in "Chiasmi International", n. 1, 1999, pp. 57-63.

Watson J.R., *The Ego and the Other*, "Tijdschrift voor Filosofie", n. 38, 1976, pp. 574-601.

Watson S., *Pretexts. Language, Perception and the Cogito in Merleau- Ponty's Thought*, "Research in Philosophy", n.10, pp. 1980, 142-166.

Watson S., *Cancellations. Notes on Merleau-Ponty's Standing between Hegel and Husserl*, "Research in Phenomenology", n. 17, 1987, pp. 191-209.

Wertheimer, M., *Productive Thinking*, Cham, Springer, 2020.

Wertz F.J., *Cognitive Psychology and the Understanding of Perception* (M. Merleau-Ponty), "The Journal of Phenomenological Psychology", n. 18, 1987, pp. 103-142.

Whiteside K.H., *Perspectivism and Historical Objectivity. Maurice Merleau-Ponty's Covert Debate with Raymond Aron*, "History and Theory: Studies in the Philosophy of History", n. 25, 2, 1986, pp. 132-151.

Whiteside K.H., *Merleau-Ponty and the Foundation of an Existential Politics*, Princeton University Press, Princeton 1988.

Whiteside K.H., *Universality and Violence. Merleau-Ponty, and the Moral Logic of Liberalism*, "Philosophy Today", n. 35, 4, 1991, pp. 372-389.

Wiggins O.P., *Merleau- Ponty and Piaget. An Essay in Philosophical Psychology*, "Man World", n. 12, 1979, pp. 21-34.

Wiggins O.P., *Merleau-Ponty's Phenomenological Ethics,* s Graduate Faculty Philosophy Journal, n. 10, 2. 1984-85, pp. 43-56.

Wiggins O.P., *Political Responsibility in Merleau-Ponty's Humanism and Terror,* "Man World", n. 19, 1986 pp. 275-291.

Yeo M., *Merleau-Ponty and the Triumph of Ambiguity*, "Semiotica", n. 85, 1-2, 1991, pp.141-146.

Zamboni C., *Ermeneutica della scrittura o iniziazione attraverso il testo?*, "aut aut", n. 202-203, pp. 1984, pp.161-171.

Zamboni C., *Note sul tema dell'immaginazione in "Il visibile e l'"invisibile" di Merleau-Ponty,* in VV.AA., *Itinerari e prospettive del personalismo*, Istituto Propaganda Libraria, Milano 1986, pp. 499-510.

Zamboni, C., *Il linguaggio nella riflessione di Merleau-Ponty e i legami con lo strutturalismo*, "aut-aut", 1989, n. 232-233, pp. 17-42.

Zani, L. *Fenomenologia dell'essere in Maurice Merleau-Ponty*, *Rivista di filosofia neoscolastica*, n. 49, 1957.
Zecchi, S., *La fenomenologia dopo Husserl nella cultura contemporanea*, voll. 2, La Nuova Italia, Florence 1978.
Zhe L., *A fundamental limit of Merleau-Ponty's trascendental phenomenology. Two concepts of intentionality,* in "Chiasmi International", n. 11, 2009, pp. 133-143.

MIMESIS GROUP
www.mimesis-group.com

MIMESIS INTERNATIONAL
www.mimesisinternational.com
info@mimesisinternational.com

MIMESIS EDIZIONI
www.mimesisedizioni.it
mimesis@mimesisedizioni.it

ÉDITIONS MIMÉSIS
www.editionsmimesis.fr
info@editionsmimesis.fr

MIMESIS COMMUNICATION
www.mim-c.net

MIMESIS EU
www.mim-eu.com

www.ingramcontent.com/pod-product-compliance
Lightning Source LLC
Chambersburg PA
CBHW031336230926
51848CB00002B/264
9788869775253